387
S

INCENTIVE PUBLICATIONS    Box 12522    Nashville, Tennessee  37212

# KIDS' STUFF MATH

activities, games and ideas for the elementary classroom

by Marjorie Frank

```
QA          Frank, Marjorie.
107
.F725       Kids' stuff math
 1974

372.86 F851k
```

Copyright © 1974 by Incentive Publications. All rights reserved. Printed in Nashville, Tennessee, United States of America. No part of this publication may be reproduced, stored in a retrieval system, or transmitted, in any form or by any means, electronic, mechanical, photocopying, recording, or otherwise, without prior written permission of Incentive Publications except as noted below.

Pages bearing this symbol ( ✸ ) were designed by the authors to be used as PUPIL ACTIVITY PAGES. It is intended that copies be made of these pages for individuals or groups of pupils. Permission is hereby granted--with the purchase of one copy of KIDS' STUFF MATH to reproduce copies of any pages bearing this symbol in sufficient quantity for classroom use.

ISBN Number 0-913916-12-9
Library of Congress catalog number 74-18907

Printed in Nashville, Tennessee
United States of America

to cheri and denise

who asked questions
i couldn't answer

and

to their parents

who never let
the questions go unexplored

Published by Incentive Publications
Box 12522
Nashville, Tenn. 37212

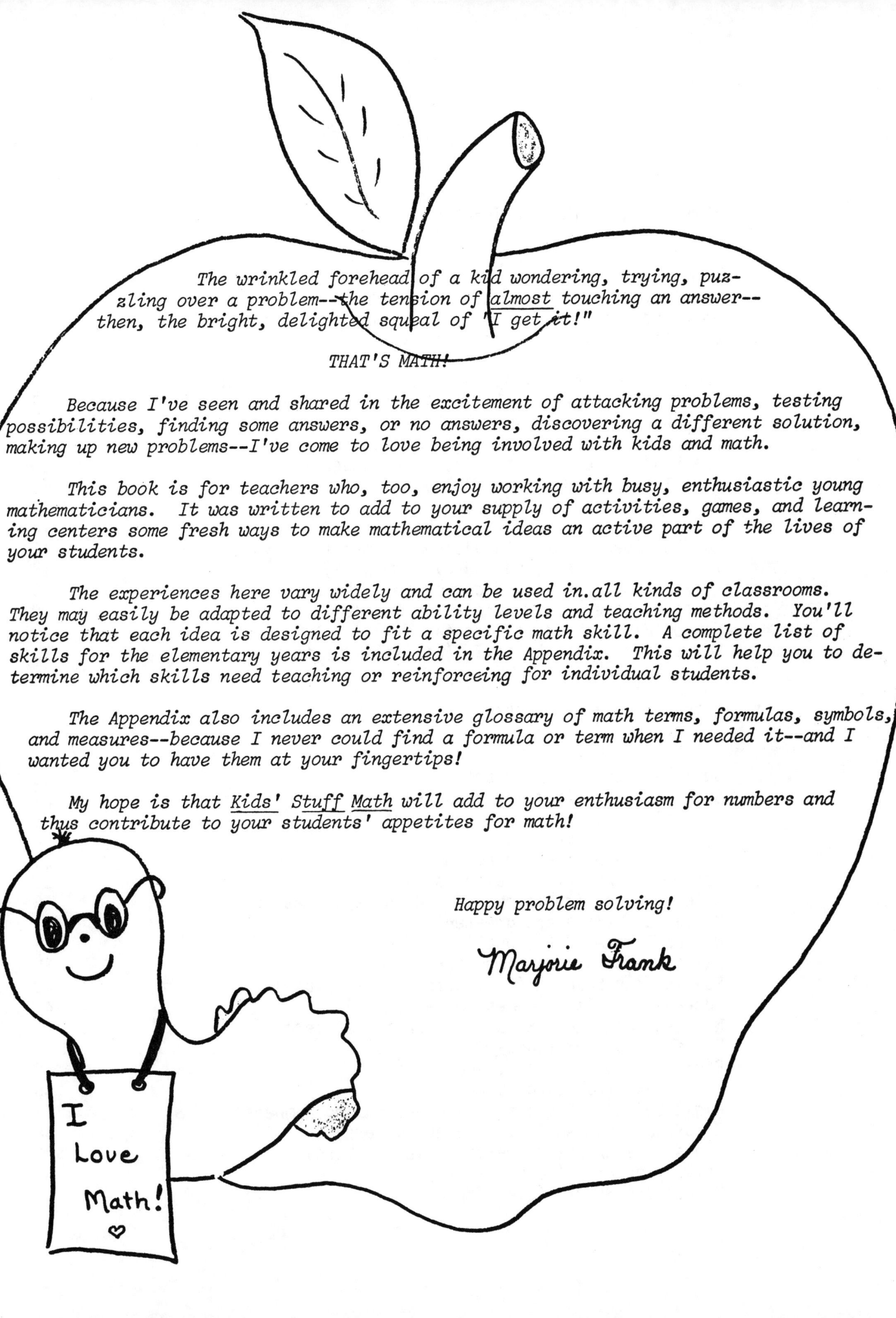

*The wrinkled forehead of a kid wondering, trying, puzzling over a problem--the tension of almost touching an answer--then, the bright, delighted squeal of "I get it!"*

THAT'S MATH!

Because I've seen and shared in the excitement of attacking problems, testing possibilities, finding some answers, or no answers, discovering a different solution, making up new problems--I've come to love being involved with kids and math.

This book is for teachers who, too, enjoy working with busy, enthusiastic young mathematicians. It was written to add to your supply of activities, games, and learning centers some fresh ways to make mathematical ideas an active part of the lives of your students.

The experiences here vary widely and can be used in all kinds of classrooms. They may easily be adapted to different ability levels and teaching methods. You'll notice that each idea is designed to fit a specific math skill. A complete list of skills for the elementary years is included in the Appendix. This will help you to determine which skills need teaching or reinforceing for individual students.

The Appendix also includes an extensive glossary of math terms, formulas, symbols, and measures--because I never could find a formula or term when I needed it--and I wanted you to have them at your fingertips!

My hope is that <u>Kids' Stuff Math</u> will add to your enthusiasm for numbers and thus contribute to <u>your</u> students' appetites for math!

Happy problem solving!

*Marjorie Frank*

# TABLE OF CONTENTS

## I. NUMERATION AND NUMBER THEORY

| | |
|---|---|
| Numo-Prints  (reading numerals) | 1 |
| What's Your Hangup?  (reading numerals) | 2 |
| Show and Tell, Chinese Style  (reading numerals) | 3 |
| Daisies Do Tell  (naming numbers) | 4 |
| All Aboard  (expanded notation) | 5 |
| Numbers Are for Navigators  (reading and writing numerals) | 6 |
| Red Hot Rod  (place value in large numerals) | 7 |
| Sock Hop  (reading numerals) | 8 |
| Does the Shoe Fit?  (naming numbers with words) | 9 |
| One Hundred Legs to Go!  (using a number line) | 10 |
| Pick Tricks!  (Roman numerals) | 12 |
| Do as the Romans Do  (Roman numerals) | 13 |
| A Case of Base  (non-decimal bases) | 14 |
| Can You Speak Zonx?  (devising a numeration system) | 16 |
| The Spider's Web  (odd and even numbers) | 18 |
| Prime Probe  (prime and composite numbers) | 19 |
| Kangaroo Capers  (factors and multiples) | 20 |
| Factor Family Tree  (prime factors) | 21 |
| How's Your Divis-Ability?  (divisibility tests) | 23 |

## II. SETS AND NUMBER CONCEPTS

| | |
|---|---|
| Search-A-Set  (forming sets) | 25 |
| Seek-A-Subset  (forming subsets) | 27 |
| Settin' Pretty  (set union and intersection) | 28 |
| The Soda Shop  (Cartesian sets) | 30 |
| Circle Set-Up  (Venn diagrams) | 31 |
| Set Membership  (equivalent sets) | 33 |
| Able Anglers  (replacement and solution sets) | 35 |
| On Target  (less and greater) | 37 |
| Winner Take All  (inequality symbols) | 38 |
| Balloon Bust  (many names for a number) | 39 |
| Say It With Shapes  (many names for a number) | 40 |
| Code-A-Note  (many names for a number) | 41 |
| Nielsen Rates Zero High  (properties of zero) | 42 |
| Please Equalize  (comparing numbers) | 43 |
| Pick a Pak O' People  (sequencing) | 45 |
| Star Trek  (rounding numbers) | 46 |
| Digit Switch  (increasing and decreasing numbers) | 47 |
| Classroom Round-Off  (rounding numbers) | 48 |
| Expose the Exponent  (exponential numbers) | 49 |
| Gallery of Art  (number ideas) | 50 |
| No Numbers Here!  (uses for numbers) | 51 |

## III. WHOLE NUMBERS AND INTEGERS

| | |
|---|---|
| Mathemat  (sums and differences) | 53 |
| Add A Row  (basic sums) | 54 |
| Finders Key-pers  (sums and differences) | 55 |
| Mr. Nobody  (identity element for addition) | 56 |
| Worm Wisdom  (associative and commutative properties) | 57 |
| Crack-A-Code  (missing addends) | 58 |
| Add and Eat  (adding several numbers) | 59 |
| Gastronomic Guesstimation  (estimating) | 61 |
| Sum Stairs  (addition and subtraction) | 62 |
| See-Saw Sums  (finding sums) | 63 |
| Palindrome Party  (patterns in addition) | 64 |
| Subtraction 500  (subtraction with renaming) | 65 |
| Tattle Tapes  (column addition) | 66 |
| Batter Up  (renaming) | 67 |
| Blockbuster  (subtraction with zeros) | 68 |
| Ways With Arrays  (writing equations) | 69 |
| Drillstrips  (addition and subtraction word problems) | 71 |
| It's A Mod, Mod World  (modular math) | 72 |
| Grid Special  (understanding multiplication and division) | 74 |
| Win-A-Pin!  (multiplication facts) | 75 |
| Hot to Dot  (missing factors and products) | 76 |
| Pail O' Problems  (related equations) | 78 |
| Grab Bag  (multiplication and division word problems) | 79 |
| Patchwork Snake  (estimating) | 80 |
| Term Squirm  (using math terms) | 81 |
| Ladybug Leap  (number line multiplication) | 83 |
| Just Average!  (averaging) | 84 |
| Deep Sea Division  (finding quotients) | 85 |
| Roll A Sentence  (using four operations) | 86 |
| Outcasts  (casting out nines) | 87 |
| Some Like It Hot . . .  (reading integers) | 88 |
| Off to the Races  (adding and subtracting integers) | 89 |
| Football Frenzy  (adding and subtracting integers) | 90 |

### *PRACTICE PAGES*

| | |
|---|---|
| No Monkey Business  (addition) | 93 |
| Magic Squares  (addition) | 95 |
| Numerals Nelda 'N Norman  (addition) | 96 |
| A-Maze-ing!  (column addition) | 97 |
| Search 'N Circle  (addition and subtraction facts) | 98 |
| Halloween + Easter = St. Valentine's Day  (addition, subtraction) | 99 |
| Tug-of-War  (basic operations) | 100 |
| Dial-A-Doughnut  (multiplication and division) | 101 |
| Napier's Bones  (multiplication facts) | 102 |
| Keep in Shape!  (multiplication facts) | 104 |
| In Living Color  (multiplication and division facts) | 105 |
| Lattice Work  (long multiplication) | 106 |
| Touch Puzzles  (all basic facts) | 107 |

## IV. FRACTIONAL NUMBERS

| | |
|---|---|
| String-A-Fraction  (fractions as parts of sets) | 109 |
| Fractions On File  (fractions as parts of sets) | 110 |
| Carton Calculators  (adding and subtracting fractions) | 111 |
| Concentration  (equivalent fractions) | 112 |
| Beau Match  (equivalent fractions) | 114 |
| Think Bunk  (equivalent fractions) | 115 |
| Common Cupboard  (greatest common factors) | 116 |
| Follow the Fraction Path  (lowest terms) | 117 |
| Caterpillars Turn into Butterflies, You Know! (renaming fractions) | 118 |
| Domin-Knows!  (adding and subtracting fractions) | 119 |
| Balance A Barbell  (mixed numerals) | 120 |
| On Your Mark  (adding and subtracting mixed numerals) | 121 |
| Fraction Frolic  (ordering fractions) | 122 |
| Easy Bull's-Eye  (multiplying fractions) | 123 |
| Boil, Bake or Fry a Fraction  (multiplying fractions) | 124 |
| Equation Race  (multiplying fractions) | 125 |
| Upside Down Makes it Right  (dividing fractions) | 126 |
| Write-A-Ratio  (writing ratios) | 127 |
| Think Delicious!  (proportions) | 128 |
| Pin-on-Proboscis  (naming ratios as percentages) | 129 |
| Per-Sense  (finding percentages) | 130 |
| Dial-A-Decimal Duo  (using decimals) | 131 |

## V. PROBLEM SOLVING

| | |
|---|---|
| Center on Solution  (creative problem solving) | 133 |
| Picture That  (picture graph problems) | 134 |
| Smarty Cat Cubby  (brain teasers) | 135 |
| Problem Pig  (word problems) | 136 |
| Bib Math  (original problems) | 137 |
| Shopping Spree  (writing equations) | 138 |
| Hey, Stack!  (original problems) | 139 |
| Make a Fast Thousand!  (using many operations) | 140 |
| Corner to Corner  (calendar problems) | 141 |
| Say It With Coins  (money problems) | 143 |
| Sport Shop  (money problems) | 144 |
| Need It or Not  (essential information) | 145 |
| Which Witch?  (determining the operation) | 146 |
| Soma This, Soma That  (verbal solutions) | 147 |
| Bug-A-Boo  (problem solving with accuracy) | 149 |
| Are You Quick on the Draw?  (multiple solutions) | 150 |
| Pass the Basket  (multiple solutions) | 151 |
| What's the Rule?  (function rules) | 152 |
| Twenty Questions  (finding unknown numbers) | 153 |
| Balance Box  (finding variables) | 154 |
| Tell Me About the Birds and the Bees!  (finding two variables) | 155 |

VI. MEASUREMENT

    Mat Magic  (more, less, greater, smaller, few, many)    157
    Time Is Alive  (telling time)    158
    How Long Do I Have to Wait?  (measuring time)    159
    "Grandfather" Knows Best  (measuring time)    160
    Service the Jet Set  (time zones)    162
    Hot Is Hot  (measuring temperature)    163
    Worm Your Way  (measuring length)    164
    Mystery Message  (measuring length)    165
    Scavenger Hunt  (measuring length)    166
    Perimeter Probe  (finding perimeter)    169
    Curve Collage  (finding perimeter)    170
    Was Pythagoras "Short"-Sighted?  (Pythagorean Theorem)    172
    Treasure Measure  (distances on a map)    173
    Where Am I?  (latitude and longitude)    175
    Anglers' Paradise  (measuring angles)    177
    Weight Watchers  (measuring weight)    178
    Let's Play Post Office  (measuring weight)    179
    Search the Area  (finding area)    180
    How Do You Figure This Figure?  (finding area)    181
    Bored?  Try Geoboards!  (finding area)    182
    Circle Circus  (measurements in circles)    183
    Figure Foil  (finding surface area)    184
    Volume Venture  (finding volume)    185
    A Gallon Is a Gallon Is a Gallon  (liquid measurement)    186
    Fill 'Em Up!  (liquid measurement)    187
    Lollipops I Can Lick  (choosing a unit of measure)    188
    Judge the Junk  (estimating and comparing)    189

VII. GEOMETRY

*Plane Geometry*

    Count and Color  (triangles and rectangles)    191
    Triple Inventory  (circles, triangles and rectangles)    193
    Anybody Home?  (closed curves)    194
    Curvaceous . . . Gracious!  (closed curves)    195
    Try Angles  (kinds of triangles)    196
    Who's Who in Quadrilaterals  (kinds of quadrilaterals)    198
    Go Geomat Style  (recognizing plane figures)    200
    Sort Sport  (classifying plane figures)    202
    Kaleidoscope  (circles)    203
    Circe-File  (circles)    205
    Geo-Bingo  (recognizing plane figures)    206
    Stitch Witchery  (plane figures)    209
    Try A Tangram  (plane figures)    210
    Go Geoboards or Bust!  (plane figures)    211
    Grid Workout  (perimeters and areas of polygons)    213
    Match Game  (recognizing polygons)    215

| | |
|---|---|
| These Angles Have Class! (identifying angles) | 217 |
| Check Your Angle (identifying angles) | 218 |
| What's Dot? (congruent figures) | 219 |
| A Bit O'Bisecting (bisecting) | 220 |
| Stick Out Your Neck--Spot Check (identifying many figures) | 223 |

### *Space Geometry*

| | |
|---|---|
| Straw Geometry (making figures) | 225 |
| Geomobiles (identifying space figures) | 226 |
| Shake, Rattle 'N Roll (identifying space figures) | 227 |
| Multi-Dip Diagnosis (identifying space figures) | 228 |
| Hard Hat Area (constructing space figures) | 229 |

### *Transformational Geometry*

| | |
|---|---|
| Symme-tree (symmetrical figures) | 236 |
| Double Up (symmetrical figures) | 237 |
| This is Symme-Rama! (symmetrical designs) | 238 |
| Slide, Flip, or Turn? (moving figures) | 240 |
| Remember Me? (recognizing figures after a move) | 242 |
| Elephant, ʇuɐɥdǝlƎ (recognizing figures in different positions) | 243 |
| Gridography (moving figures on a grid) | 245 |

## VIII. PROBABILITY, STATISTICS, and GRAPHING

| | |
|---|---|
| Chance-It! (probability) | 247 |
| How Do They Fall? (probability) | 249 |
| Jack-In-The-Box? (probability) | 251 |
| Us Behind Bars (bar graphs) | 252 |
| Day in the Round (circle graphs) | 253 |
| Lines Tell the Story (line graphs) | 254 |
| Egg Hunt (coordinate graphing) | 255 |
| Tic-Tac Graph (coordinate graphing) | 256 |
| A Happy Face at 20-20! (coordinate graphing) | 257 |
| Grid Picnic (coordinate graphing) | 258 |
| Capture the Flag (coordinate graphing) | 259 |
| Great Grid! I'm Superkid! (coordinate graphing) | 260 |
| Graph Gallery (coordinate graphing) | 261 |
| Graph-Apples (functions) | 263 |
| Line Designs (using grids) | 264 |

APPENDIX

| | | |
|---|---|---|
| I. | Skills That Make a Math Whiz | 267 |
| II. | Terrific Tables of Measures, Formulas and Symbols | 276 |
| III. | Treasures, Tools and Tidbits for Math Lovers | 280 |
| IV. | Hung Up on a Math Word?<br>(A Glossary--Every Word You Ever Wanted to Know About Math!) | 286 |

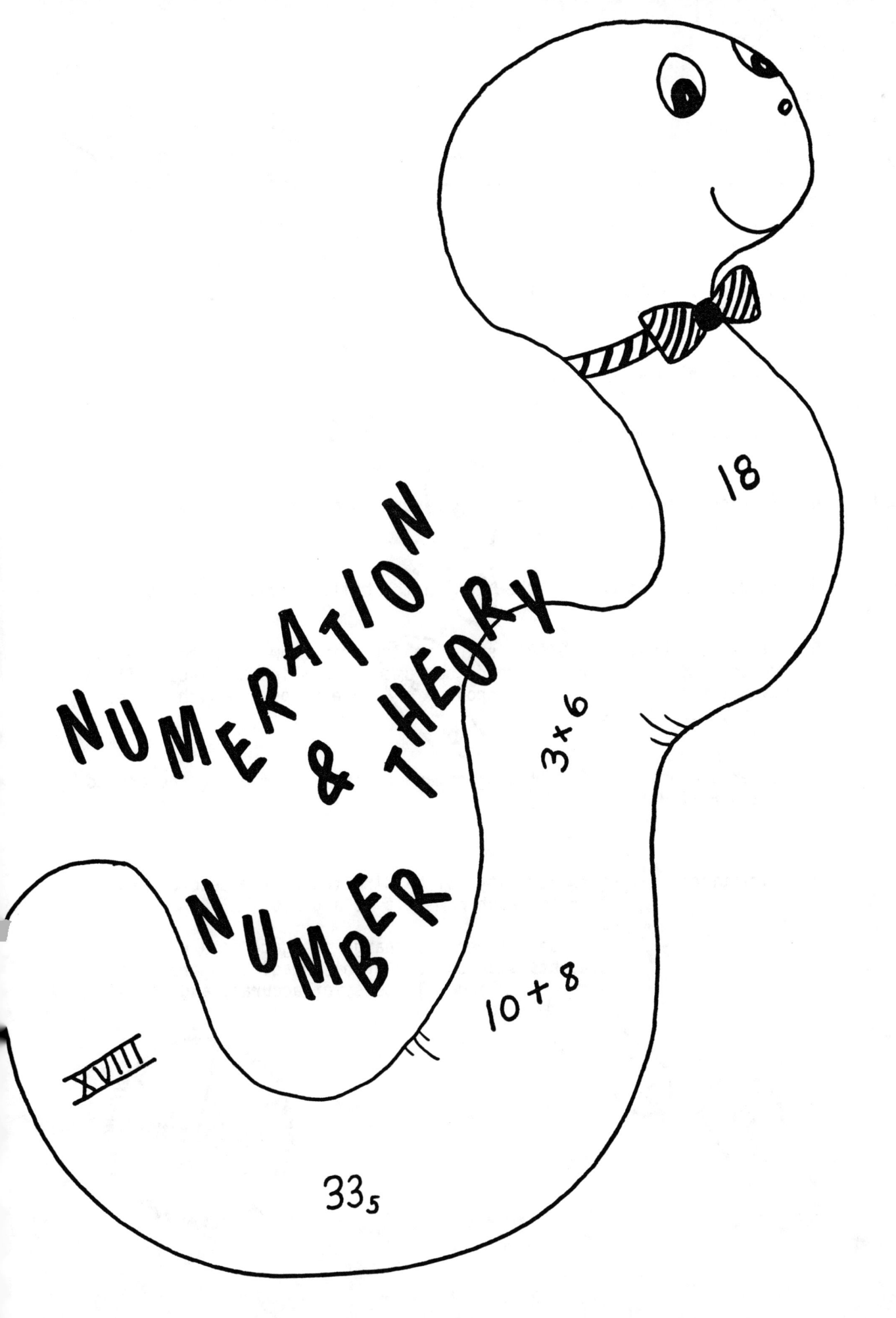

# NUMO-PRINTS

---
Skill: *Reading numerals*
---

1. Cut footprints from floor tiles, carpet scraps, or rubber carpet padding.

2. Place a numeral on each "foot."

3. Make one START and one FINISH print.

4. Provide a group of students with several numeral footprints and a spinner.

5. Students may spread out the feet in the hall, around the room, etc.

6. Player # 1 begins on START, spins to determine how many steps to advance, and must correctly read the numeral on which he lands. If he cannot read a numeral, his opponent or judge helps him with it, but he may not advance.

7. Other players follow in the same manner, the winner being the one to reach FINISH first.

   Variations:  1. This activity may be directed by the teacher and used to teach numerals to a group.

   2. Provide a timer and change the game to a race. Students step on each print, reading numerals as they go. A judge listens for accuracy and works the stopwatch!

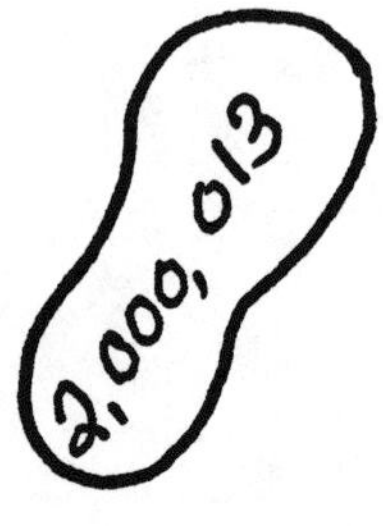

1

## WHAT'S YOUR HANGUP?

---
Skill: *Associating numerals with intervals along a number line.*
---

1. Hang a clothesline (or wire) across one end of the room (i.e., between two tables or across the bottom of a bulletin board).

2. Mark regularly spaced intervals with permanent magic marker along the rope.

3. Provide one or more sets of "socks." Each sock should bear a numeral.

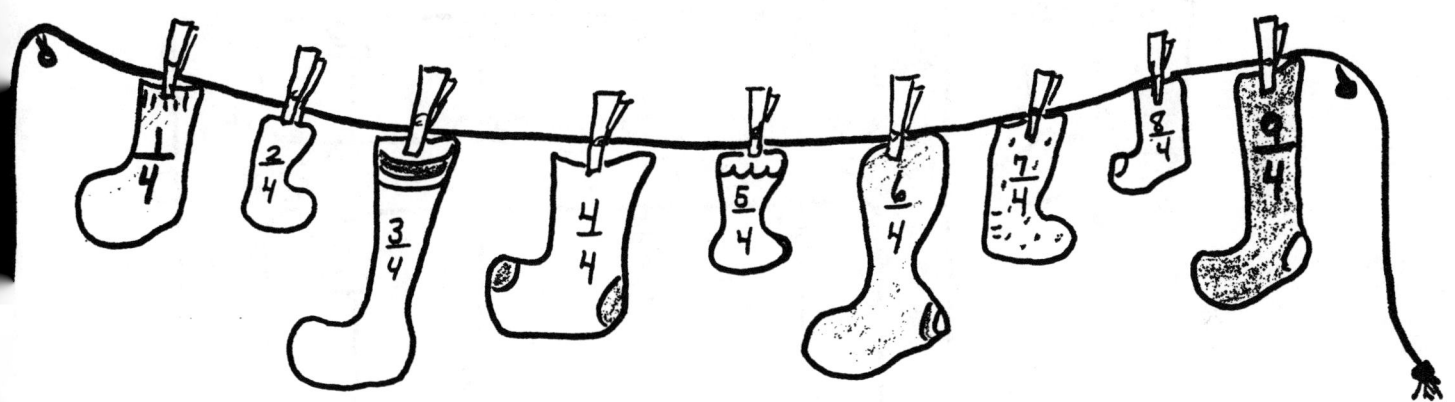

4. Provide a sack of clothespins and a sack for the "socks."

5. Ask students to order the numerals in each set and hang them at the proper intervals. (Intervals marked on the line may be treated as whole numbers, tens, etc.)

   Variation: For lower levels, socks may contain whole numbers. Decimals, integers, and fractions with unlike denominators can provide practice for more advanced students.

## SHOW AND TELL, CHINESE STYLE

---
Skill: *Reading and showing numerals using an abacus*
---

1. Make or provide an abacus for each of two or more players. The only materials needed are cardboard, heavy string, tape, and wooden beads (or macaroni rings). (See illustration.)

2. Prepare a set of cards on which you have written numerals (up to four digits) or the word names for numbers.

3. Each player takes a card, shows the number on his abacus, and reads it aloud. (He receives 1 point for a two-digit numeral, 2 points for a three-digit numeral and so on.)

4. Play continues until all the cards are gone. The player with the most points is the winner.

   NOTE: Instructions on how to use an abacus are found in the Appendix.

# DAISIES DO TELL

---
Skill: *Naming a number in several ways*
---

1. Give each student a large circle containing a number word.

2. Instruct students to make petals for their flowers. Each petal must contain another way of showing the number in the center.

3. Students should add as many petals as they can to their flower.

Variations:
1. Use larger numbers for more advanced students.

2. Make this idea into a bulletin board by "planting" several daisy centers. Put into an envelope petals with various names for each number. Students may attach these petals to the proper "centers" and add some of their own.

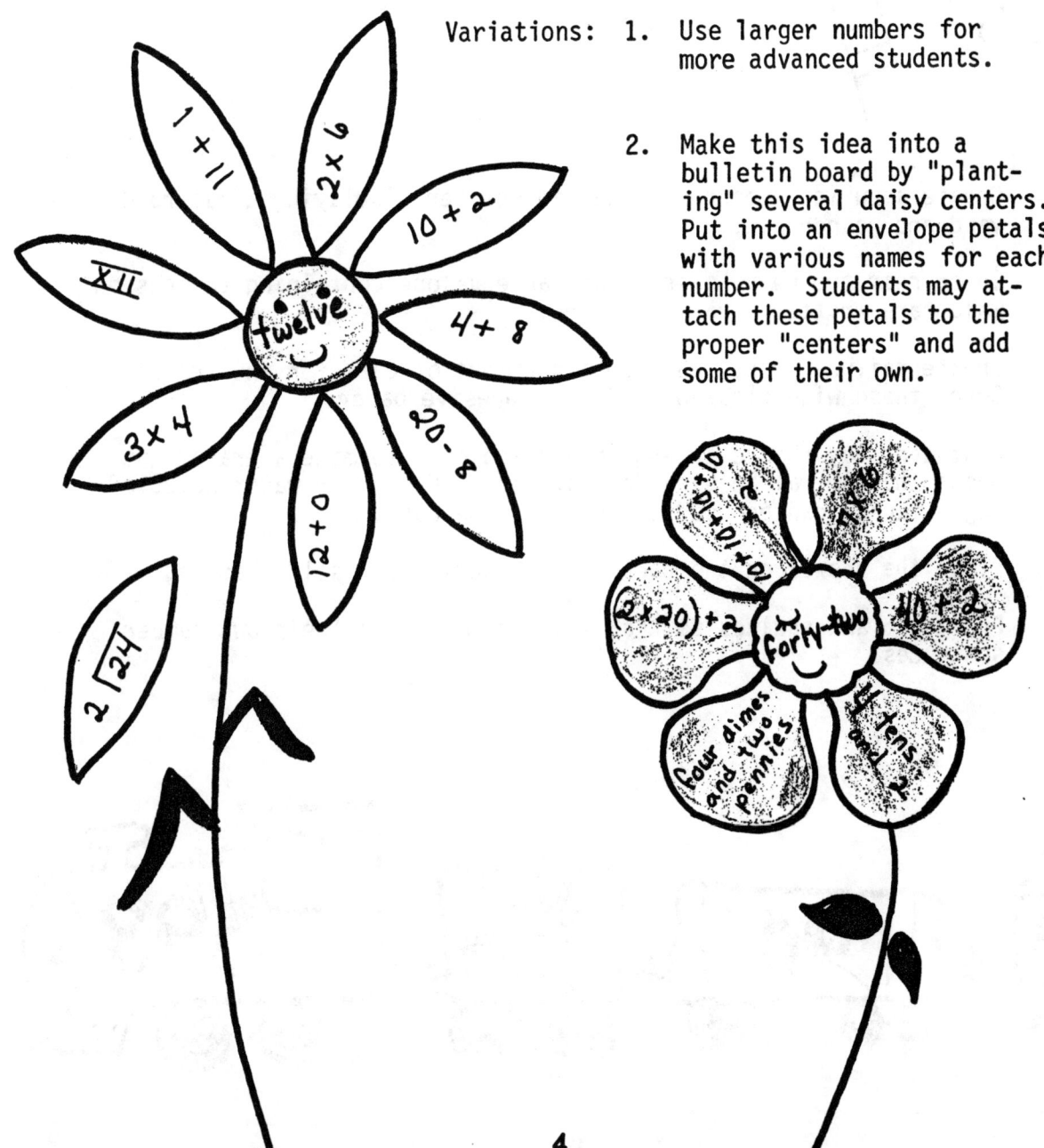

## ALL ABOARD

Skill: *Expressing numerals in expanded notation*

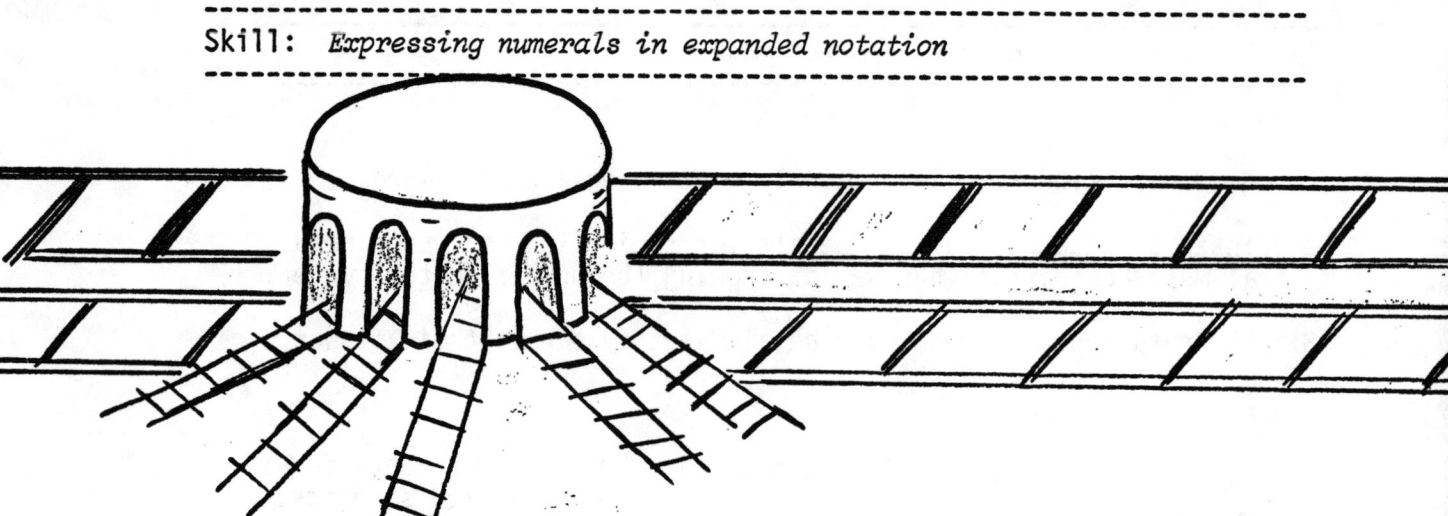

1. Create a bulletin board or learning center displaying a railroad yard and roundhouse.

2. In or near the roundhouse, place an envelope containing engines on which are written a variety of numerals.

3. In the railroad yard, park an assortment of cars for the trains. Cover these with clear plastic or adhesive paper.

4. A student may choose an engine and add cars to make a train that shows the numeral in expanded notation. Crayon or water soluble magic marker can be used to write on the cars.

5. Place the answer on the back of each engine for self checking.

6. Provide blank engines so that students may make their own number sentences.

NUMBERS ARE FOR NAVIGATORS!

---
Skill: *Reading and writing large numerals*

---

1. Prepare a center which resembles an airplane cockpit. Students will enjoy helping to create the cockpit. Install one large odometer similar to the one pictured below. (The mileage may be changed often by substituting new sets of cards or remixing the original set.)

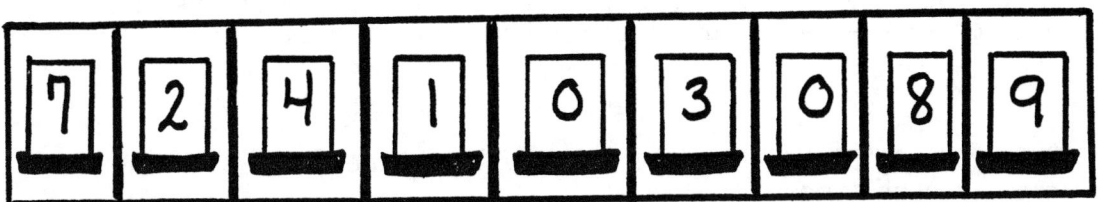

2. A pair or group of students may act as a navigation team, making and reading long "mileages" together.

3. Prepare several smaller odometers. Make them from plastic-covered tagboard so that students may write on them and erase.

4. Supply a list of instructions on tape similar to the following:

    *"Pretend that you are the pilot of a large aircraft. Your jet has traveled seventeen thousand, three hundred twenty-five miles. Show that mileage on your odometer--seventeen thousand, three hundred twenty-five miles."* (Pause) *"Did you write 1 7 comma 3 2 5?"*

    *"Now show one hundred two thousand, four hundred ninety-three miles."* (Repeat) *"Did you write 1 0 2 comma 4 9 3? I hope you remembered to write a zero in that ten thousands place."*

    *"Show this mileage: 4-6-3, 5-9-2, 0-7-1. Now read the numeral you have written."* (Pause) *"Say it again with me: four hundred sixty-three million, five hundred ninety-two thousand, seventy-one."*

    (The recorder may be disguised as a "radio" if desired! Students may add altimeters and other flight instruments displaying numbers which may be changed and read.)

# RED HOT ROD

**Skill:** *Recognizing place value in large numerals*

1. Have students draw a parking lot with 9 parking spaces ( or 12 if practice with billions is desired).

2. Also make 9 cardboard cars--one for each numeral 0 through 9. Write the numerals on the cars in black, but write one red numeral of each--i.e., 8 black and 1 red of each digit.

3. One student "parks" the cars to make long numerals, using one red car each time.

4. Another student must tell the value of the red digit.

Variation: This game may also be used with decimals.

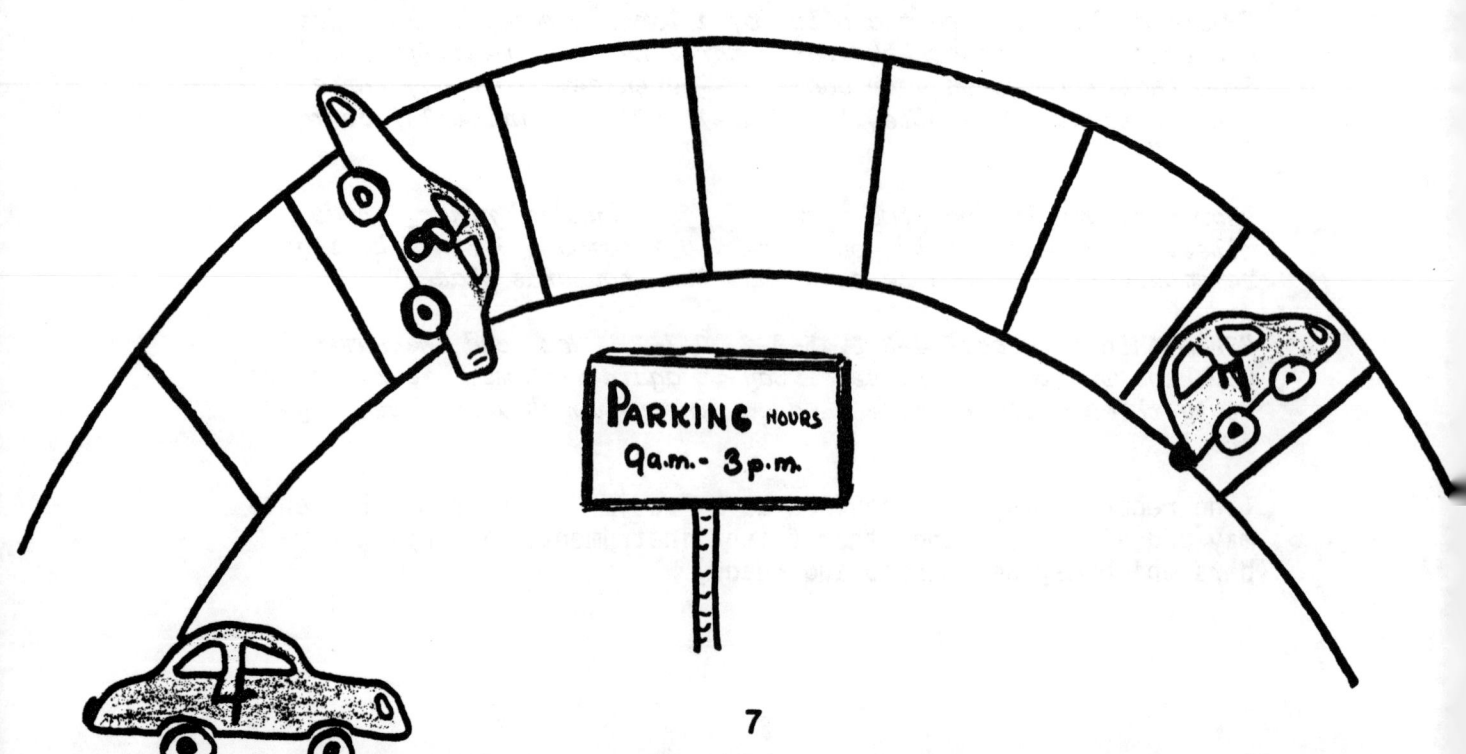

SOCK HOP

---
Skill: *Reading numerals, naming whole numbers, fractional numbers, and integers*

---

1. Make a hopscotch on a large piece of plastic (shower curtain, an old window shade, vinyl drawer liner, old plastic tablecloth, etc.).

2. On the hopscotch, write a variety of numerals.

3. Ask students to remove their shoes as they play hopscotch. They must read each numeral as they hop on it.

4. Use a liquid cleaning solution or bleach to "erase" the hopscotch and substitute new numerals.

|  | 7.8 | 1901 |
|---|---|---|
| $1\frac{7}{8}$ | .04 | -37 |
|  | 26,001 |  |
| $2\frac{4}{18}$ | 21.04 | 1.001 |
|  | $\frac{19}{2}$ |  |
|  | .003 | 14.03 |
|  | .0016 |  |

Variation: Make pieces of hopscotch from several disconnected squares so that the hopscotch will be different each time.

8

## DOES THE SHOE FIT?

---
Skill: *Associating word names with their corresponding numerals*
---

1. Cut from magazines, make, or have students draw several pairs of shoes (all kinds, sizes, shapes).

2. On one shoe of each pair, write a numeral. On the other shoe write the name of the number in words.

3. Find a container for the shoes. A dime-store hang-up shoe holder could hold several groups of shoes at various levels of difficulty.

4. Students find pairs of shoes and study the matching words and numerals.

5. Later on, make many pairs that are exactly the same, so that the type of shoe won't help in matching the pairs.

6. Be sure to provide answers--perhaps on a large shoehorn or foot!

Variation: Before school, tape a numeral or word name on the gym shoe of each student (stickers would be better). When the students arrive at school, each must find his "matching shoe."

9

ONE HUNDRED LEGS TO GO!

---
Skill: *Using a number line to read and write decimal numerals*
---

1. Make 2 centipede number lines (of different shapes), each of which is marked with tenths. Cover them with clear plastic so that students may write on them.

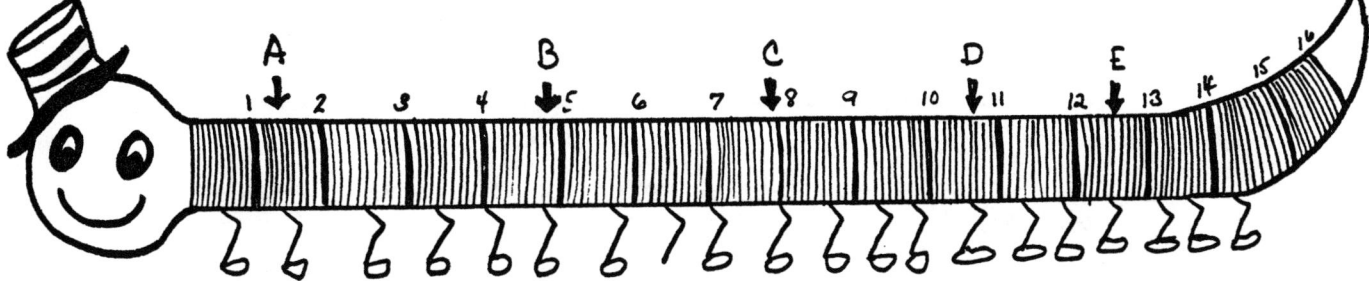

2. Along the first line place several red arrows (labeled with letters).

3. On a card, tell or write a story which includes questions such as the following (you may wish to make other cards containing more difficult questions about the same centipede):

   a. *One fair day Chester Centipede tripped on his shoelace and broke one leg. Arrow A points to the injury. Where is that leg located? Say the numeral and write it here_____.*

   b. *Chester has an ingrown toenail on the leg at Arrow D. At what location should his doctor stop to fix that problem? Read the numeral and write it here _____.*

   c. *Where is the foot that has no shoe? _____*

   and so on

   Variation: This activity might be used on an overhead projector to introduce decimals to a group. Make copies of the number lines for students so that they may work along with the teacher.

4. For the second centipede, prepare directions requiring the student to read and point out various decimal numerals. (See below.)

   a. *Sweet Cecilia Centipede is only partly numbered. You finish the numbering.*

   b. *Draw a blue shoe on the foot at 4.4. Write the numeral next to that shoe.*

   c. *Her heart is located at 2.6. Draw a heart there and write the numeral.*

   d. *Put a ribbon on her tail at 35.3. Label it with the decimal numeral.*

   *AND SO ON!*

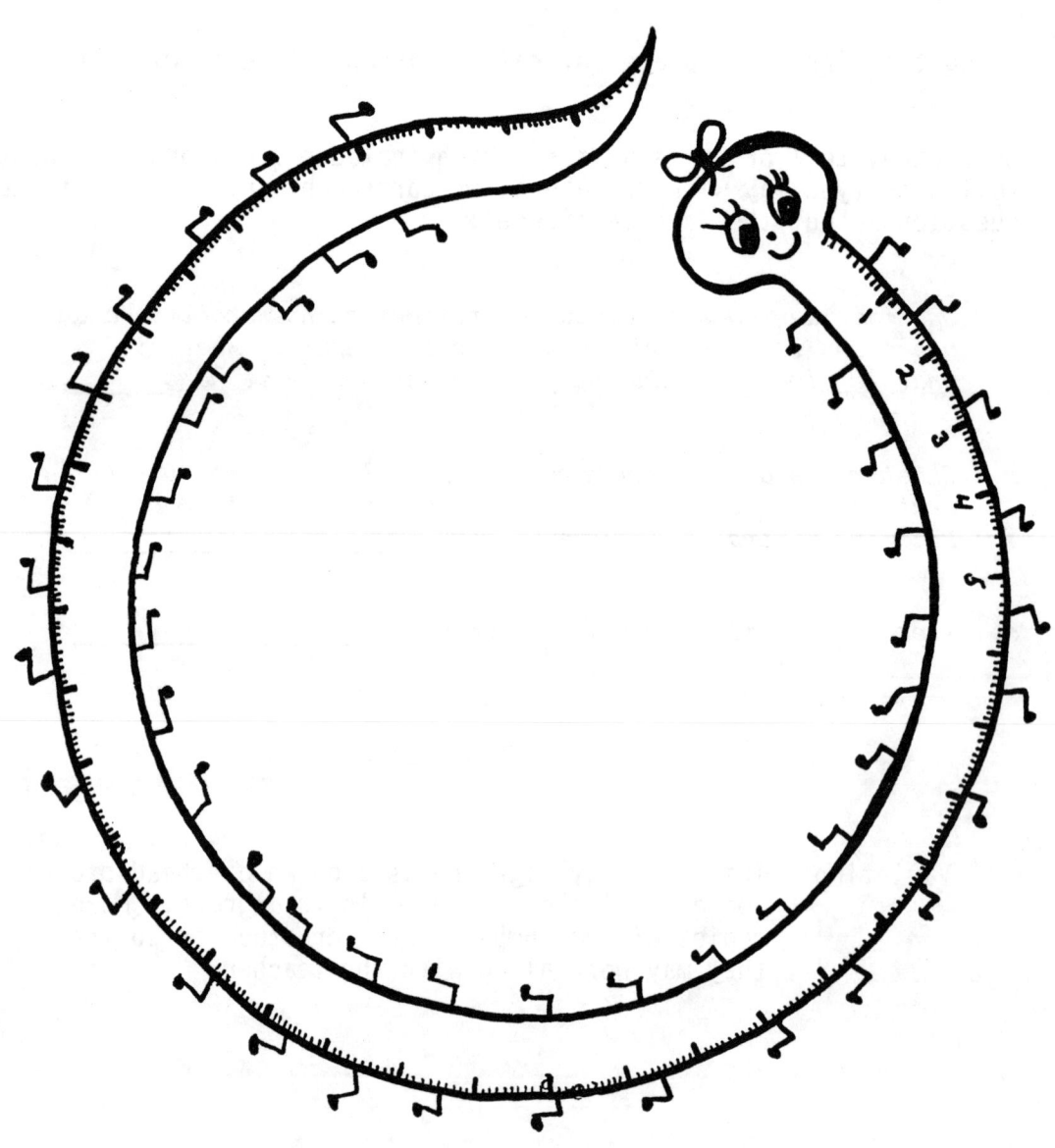

PICK TRICKS

---
Skill: *Reading and writing Roman numerals*
---

1. Provide each student with a supply of toothpicks and a large piece of construction paper on which to work.

2. Show the various Roman numerals and have students form them with their picks.

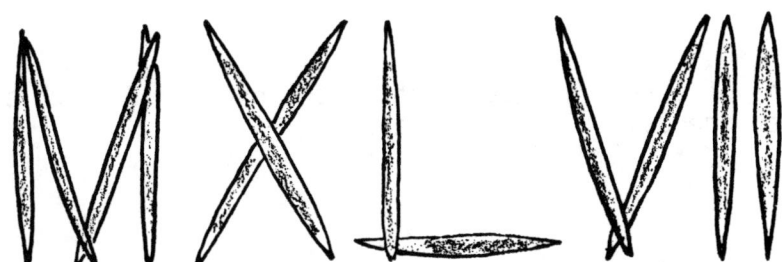

3. Have each student make a personal chart illustrating the value of each numeral by gluing the toothpicks to his paper.

4. Ask students to make larger numerals together. For example, "Show 74 at your desk," "Who can make 341?" etc.

5. Prepare a set of cards on which you have glued various Roman numerals. (Kids will want to help!) For each set, students must write the Arabic numerals. Include an answer card.

6. Provide another set of numeral cards instructing student to write each in Roman numerals with his toothpicks. Picture the numeral on the back of each card so that he may check his answers.

| | | |
|---|---|---|
| I | = | 1 |
| V | = | 5 |
| X | = | 10 |
| L | = | 50 |
| C | = | 100 |
| D | = | 500 |
| M | = | 1000 |

## DO AS THE ROMANS DO

---
Skill: *Reading and writing Roman numerals*
---

1. Make a "Roman" bulletin board, corner or center using pictures to illustrate ancient Rome.

2. Include many objects which would contain numbers (a clock, calendar, ruler, scale, thermometer, money, etc.). Label them with Roman numerals.

3. Provide copies of a record sheet such as the one here. Each student completes one using Roman numerals.

OFFICIAL RECORD BUREAU --- THE ROMAN EMPIRE

To all citizens:
You are required to supply the following information

Name _____ Age _____
Date of Birth _____
Today's Date _____
Height _____
Weight _____
Waist measurement _____
Shoe size _____
Time you go to bed _____
Your temperature _____
Amount of money you have in the bank _____
Length of your nose _____

Answer the following questions.
What is XXV + CXIV ?
What is MLVII ÷ IX ?
What is LXXIX × CCCIV ?

Suggestion: Incorporate this into a broad study of ancient Roman culture . . . including art, cooking, mythology, history, etc.

## A CASE OF BASE

---
Skill: *Reading and writing numerals written in a base other than base 10*

---

1. Make some sort of an unusual creature. Give him a large pocket which contains several digits 0 through 4 (for base 5).

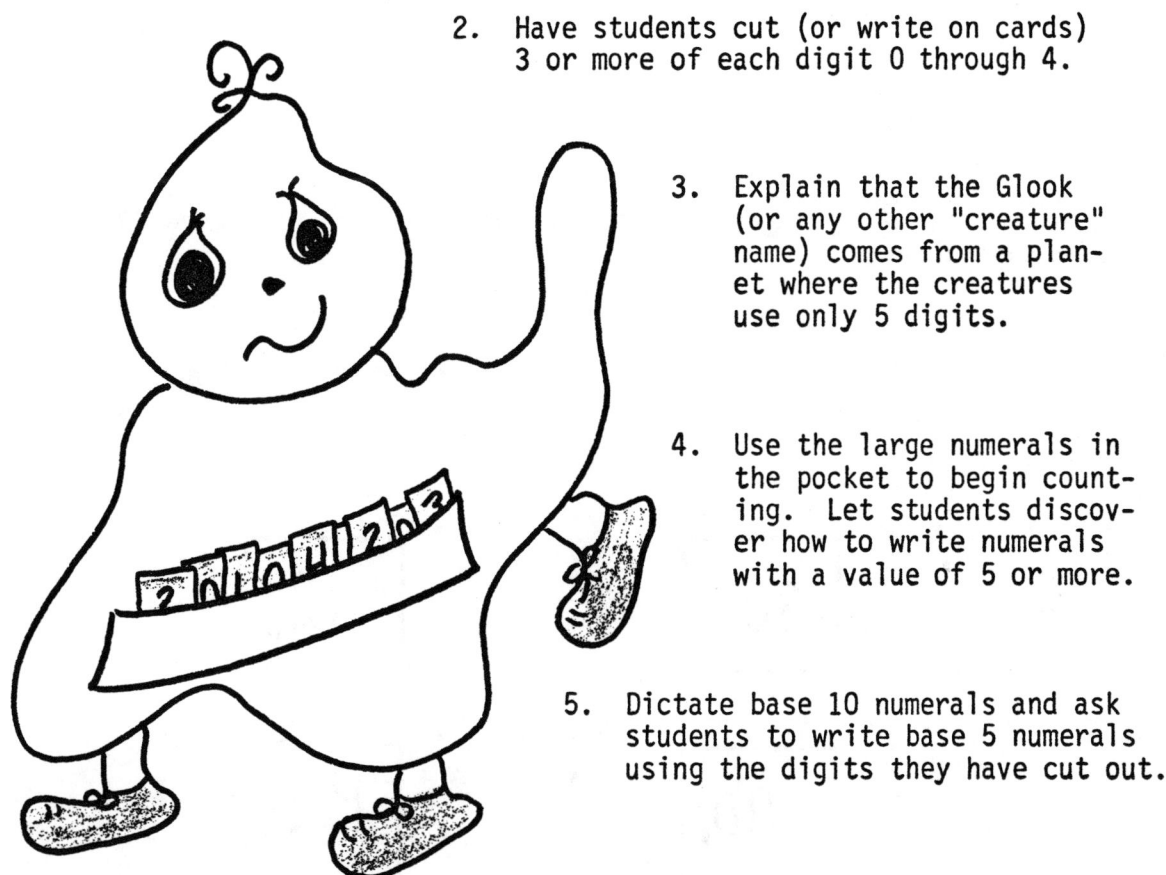

2. Have students cut (or write on cards) 3 or more of each digit 0 through 4.

3. Explain that the Glook (or any other "creature" name) comes from a planet where the creatures use only 5 digits.

4. Use the large numerals in the pocket to begin counting. Let students discover how to write numerals with a value of 5 or more.

5. Dictate base 10 numerals and ask students to write base 5 numerals using the digits they have cut out.

6. Display the Glook with some questions and activities to give students further practice converting base 10 numerals into base 5 and vice versa.

For example:

--Write your age in base 5.

--If a flea in Glookland can jump $23_5$ feet, how far could he jump here?

--Write $104_5$ in base 10.

--If Jane weighs 87 pounds, how much would she weigh in Glookland?

--Help the Glook match some base 5 numerals with your base 10 numerals.

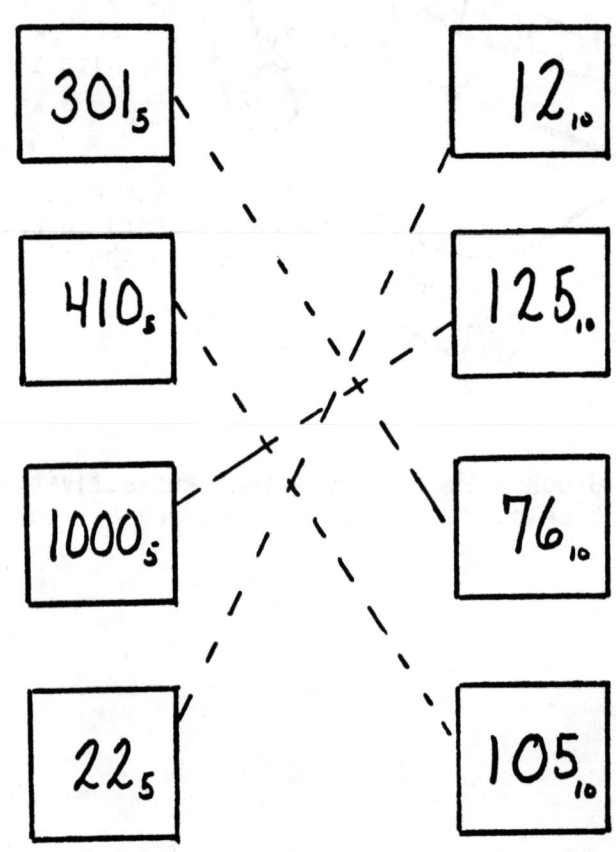

CAN YOU SPEAK ZONX?

---
Skill: *Devising and using an original numeration system*
---

1. Create an original numeration system by asking students to draw a new symbol for each of the ten digits in the decimal system.

2. Make a huge chart showing the new system. Place it in a spot where everyone may see it.

3. Practice using the new system by writing numerals together.

4. Ask each student to write (and solve) 3 problems or activities to be done using the new system. The activities may be recorded on cards and collected in a file.

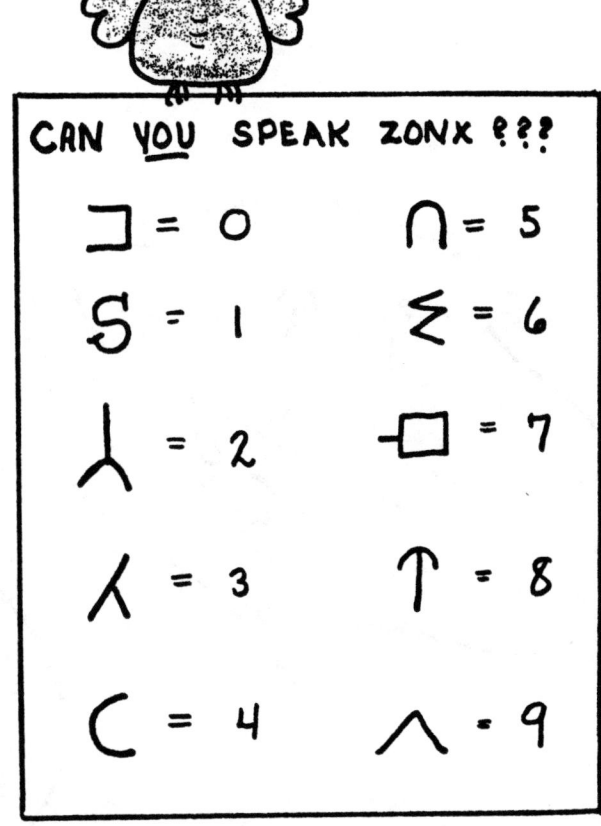

5. Do all math for one week using the new system. Use the activities suggested by students. Add your own!

   Suggestions:

   *make a time schedule for the week*
   *record weights, heights of friends*
   *write your birthday*
   *find out your mother's age*
   *paste new numerals on your clock*
   *make rulers, measuring tapes*
   *make a checkbook for yourself*
   *count calories in your new system*
   *measure and write distances to gym, lunchroom*

   Variations:
   1. Individual students may wish to make up their own private systems. Encourage them to experiment with bases other than 10.

   2. Make a scrapbook containing the different systems, or write a new book on numeration systems for your library. Each student can create problems and activities for his system.

THE SPIDER'S WEB

---
Skill: *Identifying odd and even numbers*
---

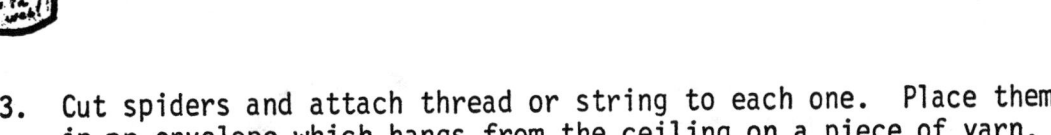

1. Make and hang two spider webs. (Use hangers, wire, or yarn stiffened by soaking in white glue.)

2. Label one web EVEN and the other ODD.

3. Cut spiders and attach thread or string to each one. Place them in an envelope which hangs from the ceiling on a piece of yarn.

4. On each spider, write an odd or even numeral or a phrase such as:

    The sum of 2 even numbers is _____.
    There is one _____ number less than 3.
    Every multiple of 2 is _____.
    The product of 2 odd numbers is _____.
    The sum of 2 odd numbers is _____.
    The product of an odd and even number is _____.

5. Students identify the numerals on the spiders or complete the sentences by hanging each spider from the correct web.

6. Include blank spiders so that each student may add one of his own to the proper web.

# PRIME PROBE

----------------------------------------------------------------
Skill: *Identifying prime and composite numbers*
----------------------------------------------------------------

1. At various places about the room, hang numbers (both prime and composite).

2. Display large clue cards which define prime numbers and tell how to spot one.

3. Have each student make a detective's notebook containing pages labeled:

    Primes Apprehended
    Composites Apprehended
    Unsolved Mysteries

4. As a student decides whether each number is prime or composite, he writes it on the proper page in his notebook. The "Unsolved Mysteries" page is for "number cases" he can't crack!

Suggestion: Put the numbers on footprints around the walls to make the search more interesting.

23    7    97

67         3

        31

## KANGAROO CAPERS

---
Skill: *Identifying factors of a number and multiples of a number*
---

1. Display two large kangaroos. Label one FACTORS and the other MULTIPLES.

2. Label letter-size envelopes with the numbers for which students need practice in identifying factors and multiples. Place these envelopes in the corresponding kangaroo's pouch.

3. Provide a large supply of blank paper strips.

4. Students write the factors which will fill each pocket on the FACTORS kangaroo.

5. They follow the same procedure for the MULTIPLES kangaroo.

6. Students may use a matrix chart of multiplication facts for checking their work.

   Variation: You fill the envelopes and ask students to decide if you have correctly identified the factors and multiples. Instruct them to keep a list of numbers that do not belong or are missing from an envelope.

## FACTOR FAMILY TREES

------------------------------------------------------------
Skill: *Identifying prime factors of a number*
------------------------------------------------------------

1. Make a poster or bulletin board showing students how to find the prime factors of a number. Use yarn or colored straws and large circles to make your factor "tree."

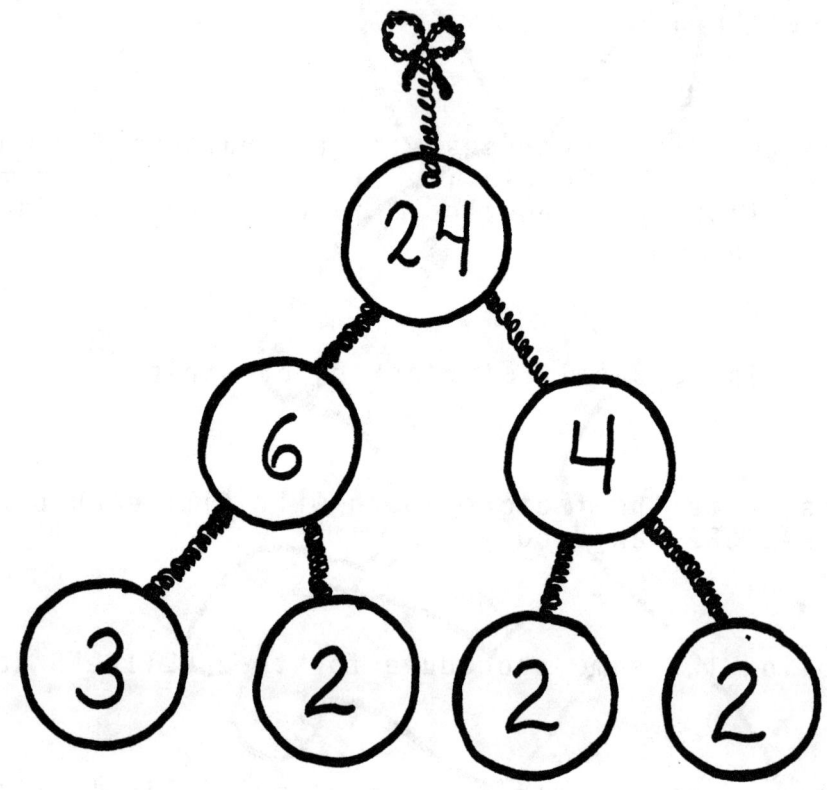

2. Make a pocket from which students can draw a number. In it, place several numbers on circles. Punch a hole near the top of each circle.

3. Provide a box of pipe cleaners and cardboard discs.

4. Students then make their own factor trees for the numbers they draw by gluing together pipe cleaners and circles.

5. Display the trees by hanging them from a wire, a cord, or clothesline.

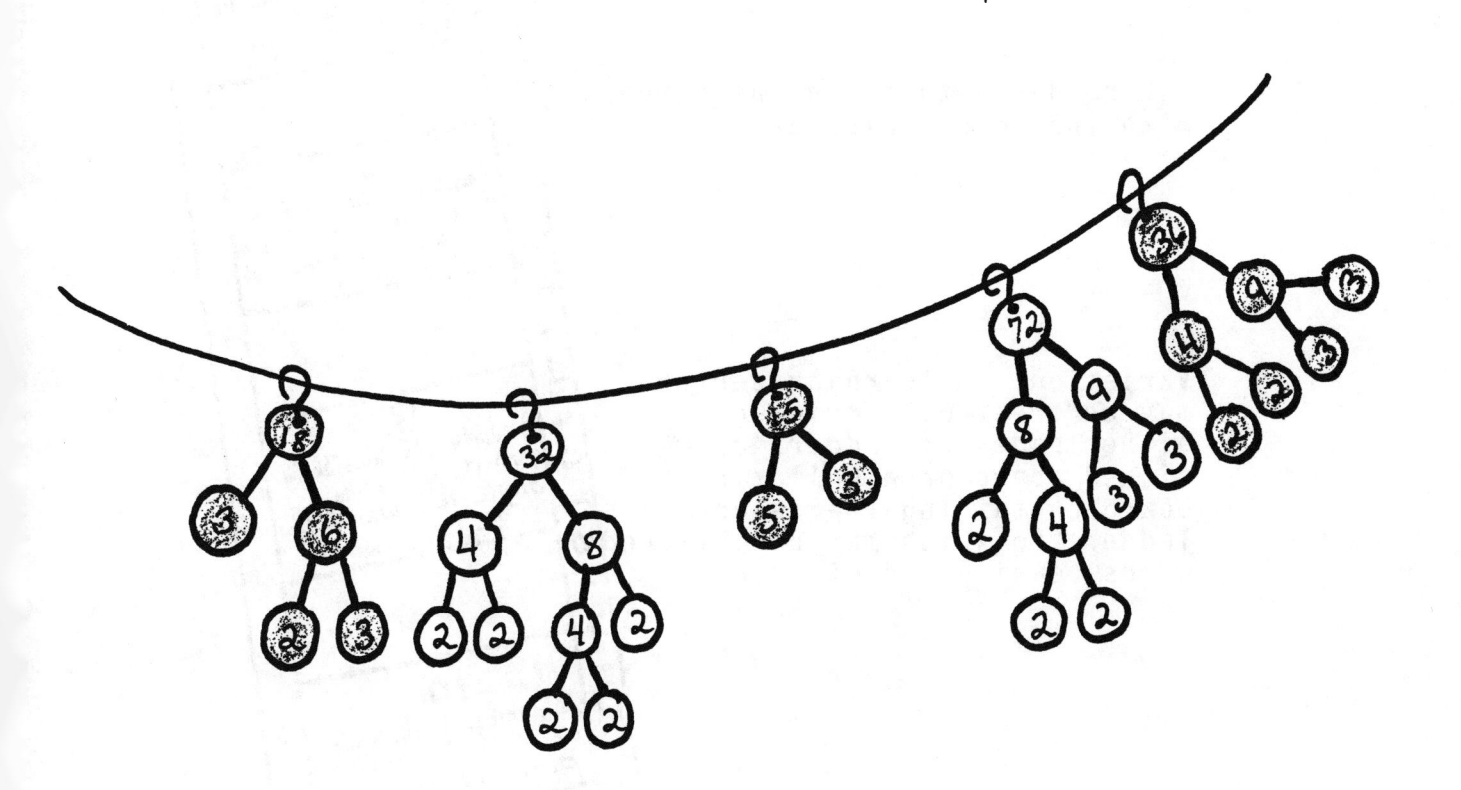

## HOW'S YOUR DIVIS-ABILITY?

---
Skill: *Determining if a whole number is divisible by 2, 3, 4, 5, 6, 9, or 10*
---

1. Make and display a large ladder which gives clues for the divisibility of whole numbers.

2. Duplicate a sheet containing 7 ladders. Label each ladder with one of the numbers 2, 3, 4, 5, 6, 9, 10.

3. On another sheet, list numbers divisible by 2, 3, 4, 5, 6, 9, or 10.

4. Ask students to match the numbers with the correct ladders.

Variation: A learning center may be created, providing either pages of ladders as mentioned above or a bulletin board containing the seven ladders to which may be attached cards bearing the dividends.

DIVISIBILITY LADDER

**2** A number is divisible by 2 if the last digit is 2, 4, 6, 8, or 0.

**3** A number is divisible by 3 if the sum of the digits is divisible by 3.

**4** A number is divisible by 4 if the last 2 digits are divisible by 4.

**5** A number is divisible by 5 if the last digit is 0 or 5.

**6** A number is divisible by 6 if it is even and the sum of its digits is divisible by 3.

**9** A number is divisible by 9 if the sum of its digits is divisible by 9.

**10** A number is divisible by 10 if the last digit is 0.

23

# SETS and NUMBER CONCEPTS

## SEARCH-A-SET

---
Skill: *Classifying objects into sets*
---

1. Have students cut and mount dozens of pictures of objects and people.

2. Mix the pictures and place them in a box.

3. Students classify objects according to a characteristic common to the group.

4. Ask them to keep a list of the sets they have formed.

5. Encourage students to begin a second time and try to form different sets.

OR

1. Give students a pile of magazines and a list of Set Titles.

   i.e:  people with curly hair          things that come in pairs
         fattening desserts              means of transportation
         hairy dogs                      vacation spots
         things that fly                 things to avoid
                                         washday products
                                         household appliances

25

2. Have each student make a booklet of five pages, headed with his choice of five titles.

3. Each page can be filled with cut-out pictures of objects which would fit into that set.

SEEK-A-SUBSET

---
Skill: *Separating members of a set into subsets*
---

1. Fill a large bag with objects belonging to one set (i.e., wooden things: blocks, ruler, pencil, jewelry, sticks, etc. OR round things: buttons, coins, rings, paper circles, embroidery hoops, plate, earring, wheel, etc.).

2. On the bag, write instructions directing the students to examine the contents of the bag and separate the set into subsets.

*Take everything out of this bag.*
*Separate the objects into subsets.*
*Make a list of your subsets.*
*Try to make some different subsets.*

3. Students may record the subsets they make by writing a list or by drawing and labeling their subsets.

## SETTIN' PRETTY

---
Skill: *Finding and naming members in the intersection or union of two sets*
---

1. Make about fifty brightly colored ☺s, 3 in. in diameter. (Snowflakes, jack-o-lanterns, bells, hearts, etc. might be substituted, depending on the season.) Place them in random pattern on a bulletin board or chart.

2. Use yarn to enclose some of the ☺s into three or four sets, such as below:

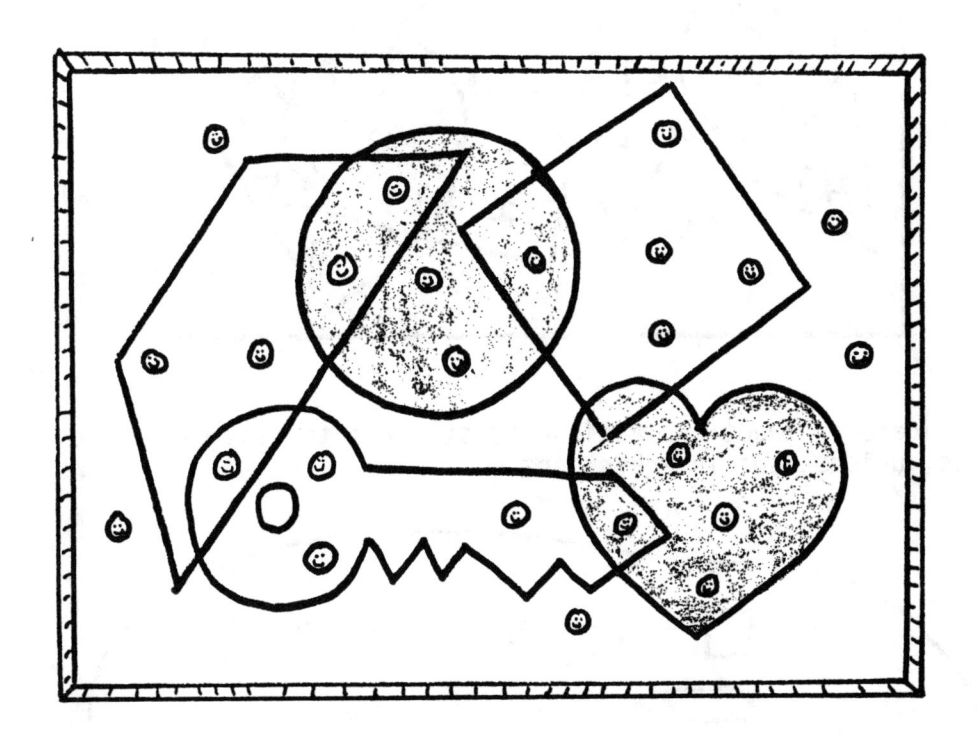

3. Ask students to tell:

   How many ☺s are in Set ♡ ?
   How many ☺s are in Set ⬠ ?
   How many ☺s are in Set ◇ ?
   How many ☺s are in Set ⚷ ?
   How many ☺s are in the union of ◇ and ○ ?
   How many ☺s are in the union of ♡ and ⚷ ?
   How many ☺s are in the intersection of       and       ?

   and so on . . .

4. Later, change the sets by making new boundaries with the yarn. Ask the same kinds of questions again.

5. Still later, remove the ☺ shapes, and substitute numerals:

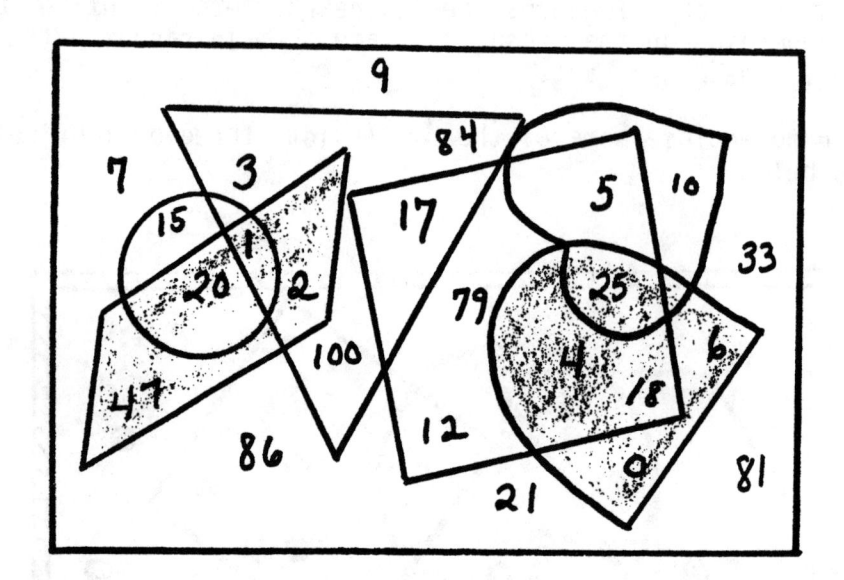

6. Duplicate instructions asking the students to list members of sets, unions, and intersections.

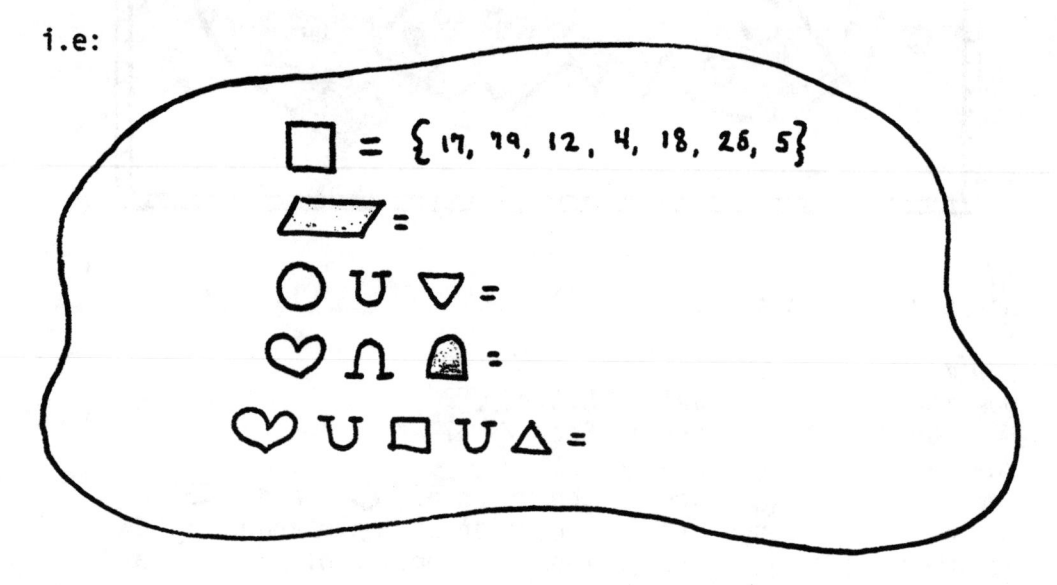

   i.e:

   □ = {17, 79, 12, 4, 18, 25, 5}
   ▱ =
   ○ ∪ ▽ =
   ♡ ∩ ⌂ =
   ♡ ∪ □ ∪ △ =

THE SODA SHOP

---
Skill: *Forming Cartesian sets*
---

1. Create a center which will resemble an ice cream parlor or soda shop. Include pictures or empty containers suggesting the following menu:

   | DRINKS | ICE CREAM | TOPPINGS | CONES |
   |---|---|---|---|
   | chocolate soda | vanilla | chocolate | sugar |
   | rootbeer float | chocolate | strawberry | regular |
   | cherry punch | peach | butterscotch | |
   | lemonade | strawberry | marshmallow | |
   | | banana | pineapple | |

2. Supply a list of questions which require students to form Cartesian sets.

   Example:

   *How many different sundaes could be made?*

   *Write them down or draw them.*
   *Write a number sentence and show the answer.*

   *How many kinds of single dip cones could you buy?*

   *List or draw them.*
   *Write a number sentence to go along with your list.*

   *If Tracy wants a dish of ice cream and a drink, how many different combinations might she choose?*

   *Name or draw them.*
   *Write a number sentence to show how many choices she has.*

   Variation: Add prices to the items and you'll have dozens of problems available for solving!

30

## CIRCLE SET-UP

---
**Skill:** *Representing sets and intersections of sets using Venn diagrams*
---

1. Prepare: --several circles of yarn (24" pieces)
   --cardboard numerals
   --small cards containing capital letters (for labeling sets)
   --a deck of Set Cards containing descriptions or members of sets:

   > i.e: *(1, 4, 7, 9, 3)*     *multiples of 3 below 30*
   > *factors of 24*     *odd numbers under 10*

   --a deck of Direction Cards:

   > i.e: *Draw 1 card.*     *Show the set.*
   > *Draw 3 cards.*     *Show the intersection of the 3 sets.*
   > *Draw 2 cards.*     *Show the union of the 2 sets.*

2. Keep all the parts of the game in a flat box. Directions may be written inside the box top.

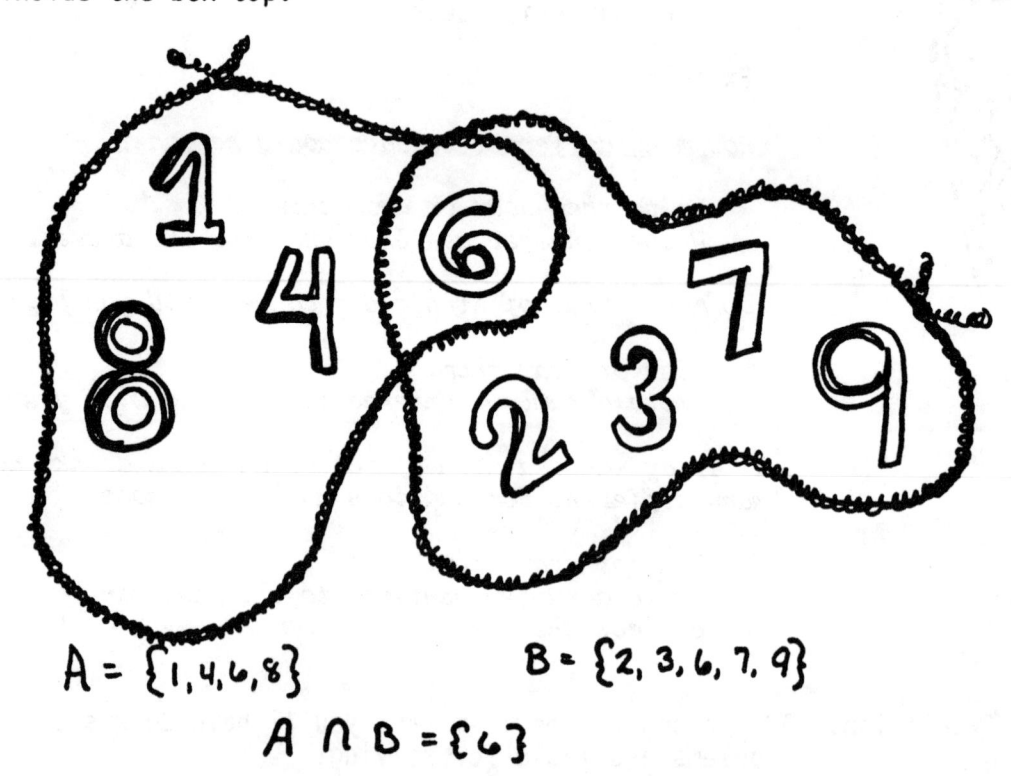

$A = \{1, 4, 6, 8\}$     $B = \{2, 3, 6, 7, 9\}$

$A \cap B = \{6\}$

3. As a group of students plays the game, each player draws a Direction Card and does what it asks. Using yarn circles and cardboard numerals, he represents the sets in Venn Diagrams, then writes or draws his results.

4. Papers may be checked by teacher or a helper to see if the players are using sets with understanding.

Variation.  1. For younger children, use pictures and objects instead of numerals.

2. You may ask students to draw the diagrams on papers, making the circles with crayon.

Start a class Number Notebook. Throughout the year, everyone can contribute new number ideas: facts, history, patterns, tricks, etc.

Let groups of students make mystery movies about numbers. After seeing a movie, the other students try to identify the "suspected" number.

Looking for objects to use as counters, or for making sets?  TRY:

| marshmallows | corn | jelly beans |
| shoes | beans | people |
| pennies | bottle caps | marbles |
| straws | pencils | rulers |
| legs | puffed rice | tiddlywinks |
| buttons | fingers | poker chips |
| paper cups | graph paper strips | dominoes |
| small blocks | beads | spoons |
| M & M's | playing cards | scissors |

SET MEMBERSHIP

---
Skill: *Listing members of sets and determining whether two sets are equivalent or non-equivalent*
---

1. Prepare a set of 20 cards (one-foot square poster board) on which have been glued sets of small objects such as corn kernals, pennies, small pictures, numerals, buttons, etc.

2. Label each card with a letter.

3. Duplicate the following instructions and give to each student.

---

**Here's What to do with Your Set Cards!**

I. List the members of each set .... like this:
   A = {penny, marshmallow, button, m+m}
   B
   C
   D
   E
   F
   G
   H
   I
   J
   K
   L
   M
   N
   O

Name _____

II. How many members are in each set?
   A           F           K
   B           G           L
   C           H           M
   D           I           N
   E           J           O

III. Match the sets that are equivalent. Which ones are?

IV. Draw a set here that is equivalent to J:

# ABLE ANGLERS

**Skill:** *Identifying replacement and solution sets for number sentences*

1. Prepare on cardboard "worms" a set of number sentences from which numbers or solutions have been omitted.

2. Make a "fish tank" such as the one below. On each fish, write a number or combination of numbers which could replace or solve the missing parts. Attach some metal object to the back of each fish, so that it may be attracted by a magnet.

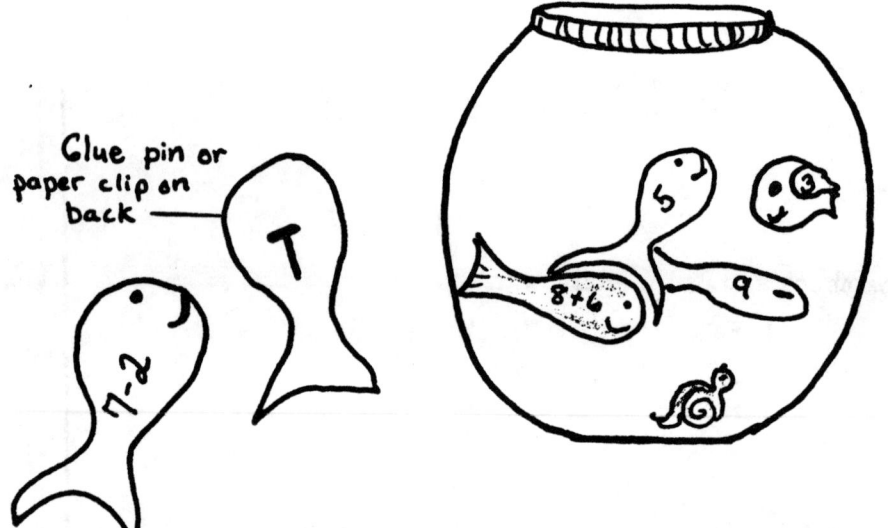

3. A fishing pole may be made by attaching a string to a stick. At the end of the string tie a small magnet.

4. Students choose a worm and must catch all the fish which belong to the replacement set or solution set for their number sentence.

5. After all the fish are caught by that worm, the student writes his answers on a duplicate of the following page.

# How's Your Fishing?

Name _____

This is Fred

This is Fred's Worm

These are the fish Fred caught. (4, 5, 3+2, 6+1)

Fred wrote his answers on the first line.
You write your answers in the same way.

| Sentence on Worm | Set that could replace ⬜n or solve the sentence | Is this a replacement set? | or a solution set? |
|---|---|---|---|
| 8 − ⬜n > 2 | {1, 4, 5, 3+2, 6+1} | yes | |

## ON TARGET

**Skill:** *Given 2 whole numbers, identifying which is less or greater*

1. Print several numbers on a large piece of cotton fabric.//
2. Mount the fabric on sturdy cardboard or an old bulletin board.
3. Supply darts and a small chalkboard for players.
4. Each player throws 2 darts. He must write a number sentence on the chalkboard showing the relationship between the 2 numbers.

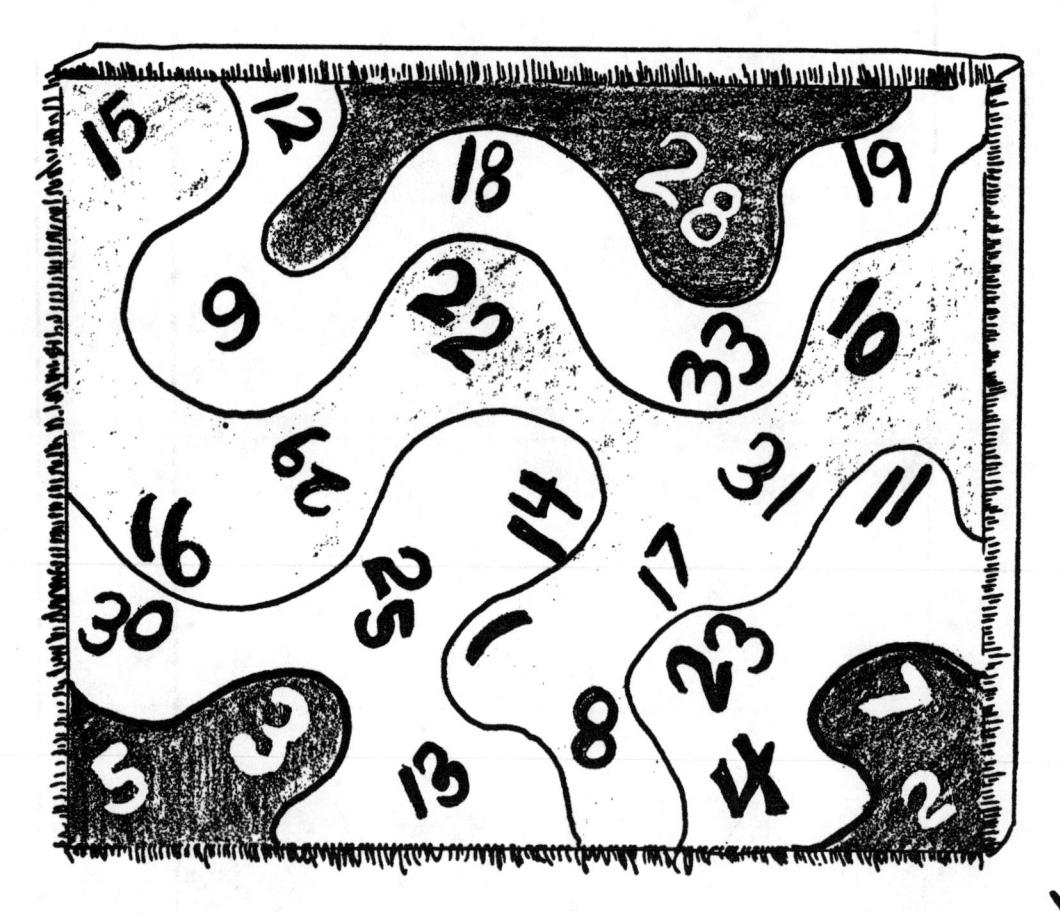

**Variation:** To make the activity more interesting and difficult, use fractions, decimals, mixed numerals, integers, equations in place of single whole numbers.

## WINNER TAKE ALL

---
Skill: *Identifying using and understanding the symbols < > = ≠*
---

1. Prepare a deck of cards containing a variety of numerals.

2. Make a second deck by printing one of the following symbols on each card:

    <  >  =  ≠

3. Players deal all the Numeral Cards.

4. As each player takes a turn, he draws one Symbol Card and lays down 3 cards to make a sentence. If he cannot make one, he must pass.

5. Players continue until someone is out of cards. The player with the most number sentences is the winner.

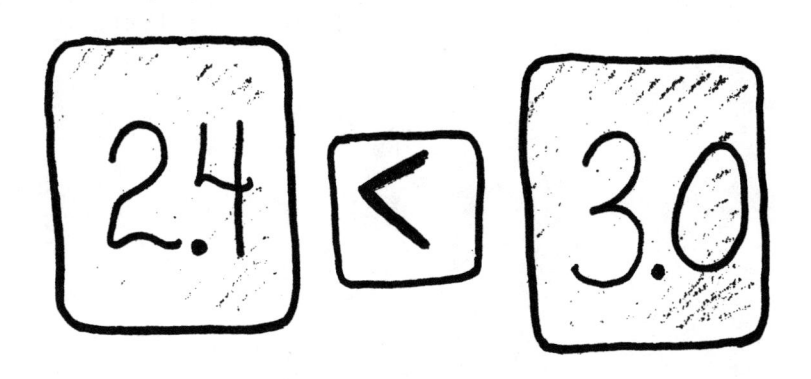

BALLOON BUST

---
Skill: *Giving many names for a number*
---

1. When each student comes into class, pin a number on his shirt.

2. Provide paper and string for making balloons.

3. During the day, or Math period, each student makes as many balloons as he can. Each one must show a way to name the number he has been given.

4. Tie together each "bunch" of balloons with a piece of yarn and a label bearing the original number.

Variation: This activity might be a contest to see who can form the largest bunch of balloons. A group of students or a whole class could work together. You may need to set a time limit.

## SAY IT WITH SHAPES

---
Skill: *Writing number sentences to name numbers in various ways*
---

Duplicate a sheet similar to the one below.

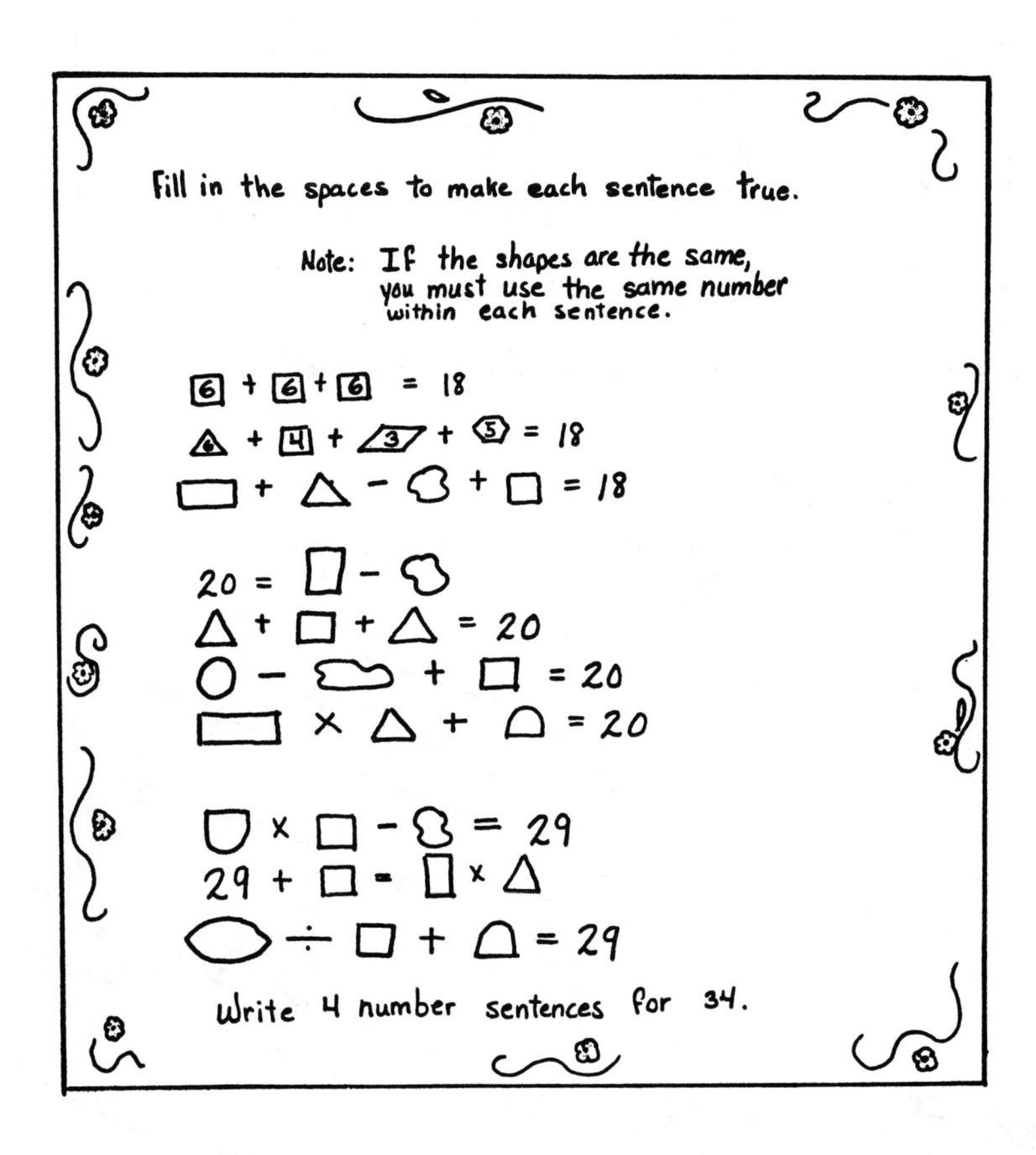

---

Fill in the spaces to make each sentence true.

Note: If the shapes are the same, you must use the same number within each sentence.

⬜6 + ⬜6 + ⬜6 = 18

△ + ⬜4 + ⟋3⟍ + ⬠5 = 18

▭ + △ − ☁ + ▢ = 18

20 = ⬜ − ☁

△ + ▢ + △ = 20

○ − ☁ + ▢ = 20

▭ × △ + ⌒ = 20

⋃ × ▢ − ☁ = 29

29 + ▢ = ⬜ × △

⬭ ÷ ▢ + ⌒ = 29

Write 4 number sentences for 34.

---

40

## CODE-A-NOTE

---
**Skill:** *Recognizing a number written in various ways*
---

1. Devise a code using numbers as symbols for letters:

   |  |  |  |  |
   |---|---|---|---|
   | 23 = A | 4 = H | 21 = O | 1 = V |
   | 2 = B | 10 = I | 18 = P | 40 = W |
   | 100 = C | 6 = J | 5 = Q | 30 = X |
   | 31 = D | 9 = K | 7 = R | 12 = Y |
   | 0 = E | 3 = L | 99 = S | 25 = Z |
   | 15 = F | 27 = M | 13 = T | |
   | 16 = G | 8 = N | 101 = U | |

2. Write a coded message to be deciphered by students.

3. Ask students to write you a note using the same code!

   **Variation:** Keep a collection of messages for duplication so that you can give out different ones for each individual.

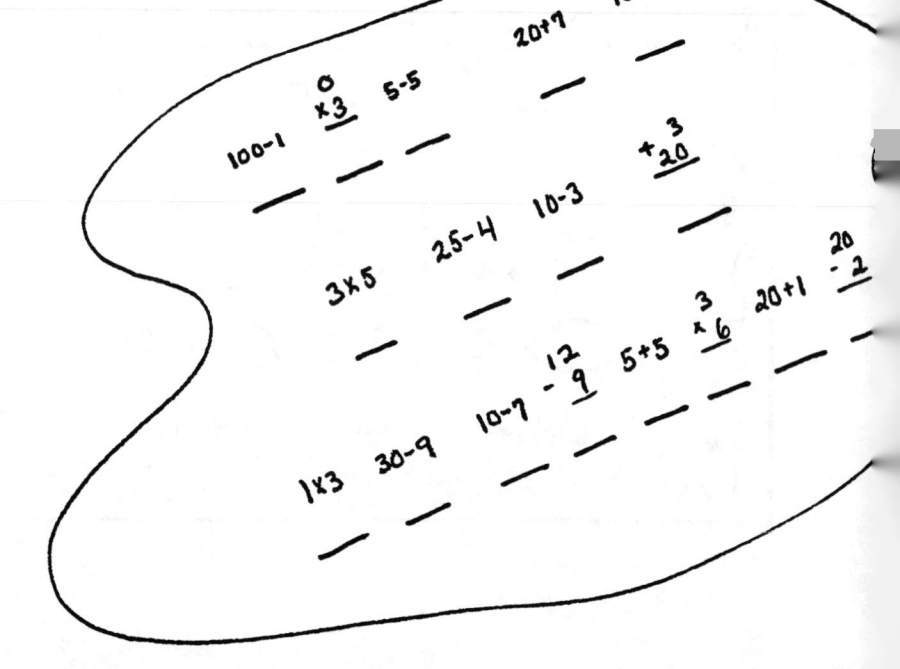

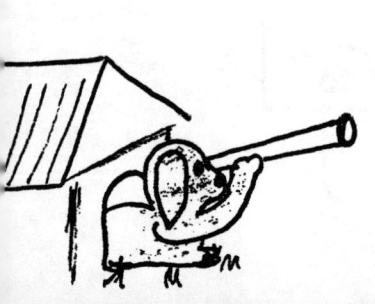

## NIELSEN RATES ZERO HIGH!

Skill: *Identifying the properties and functions of zero*

1. Find an old (empty) TV set, or make one from a cardboard box.

2. Have students gather all the information they can about the number 0 (zero) and prepare a show

    "The Adventures of Zero."

Variation: Try playlets about other numbers too!

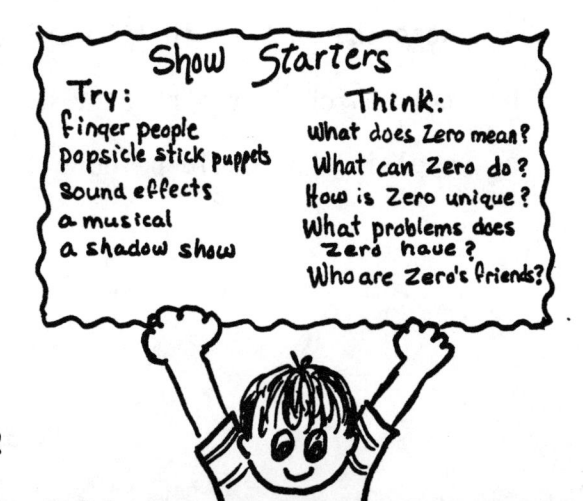

**Show Starters**

Try:
- finger people
- popsicle stick puppets
- sound effects
- a musical
- a shadow show

Think:
- What does Zero mean?
- What can Zero do?
- How is Zero unique?
- What problems does Zero have?
- Who are Zero's friends?

## PLEASE EQUALIZE

---
Skill: *Comparing and equalizing values of numbers*
---

1. On small index cards write pairs of numbers (or descriptions of numbers requiring some counting, measuring, etc.).

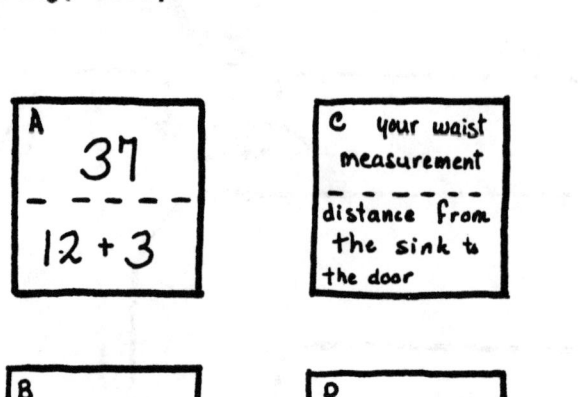

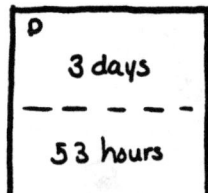

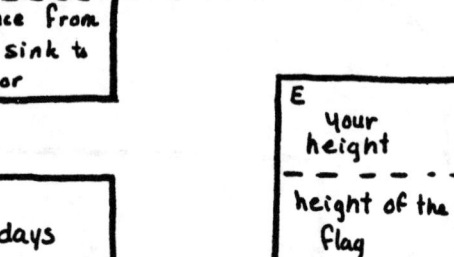

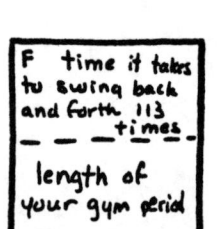

   **A** 37 / 12 + 3

   **B** 2 dozen eggs / 11 eggs

   **C** your waist measurement / distance from the sink to the door

   **D** 3 days / 53 hours

   **E** your height / height of the flag

   **F** time it takes to swing back and forth 113 times / length of your gym period

2. Give a card to each student. Allow a time period long enough for him to do any figuring, counting, or measuring.

3. Each student must be prepared to:

   a. describe numbers in like units.

   b. tell which is the greater amount.

   c. explain what could be done to equalize the values.

43

Variations:
1. Put all the cards into a "Please Equalize Us" Center. Provide a record-keeping sheet so that one student may solve several of these number puzzles. Within the center, vary the difficulty of the tasks.

2. For students at a lower level, make the problems much simpler. (See illustrations below.)

3. For concentration on one process, make all cards deal only with that one process, i.e., measurement only, multiplication only, etc.

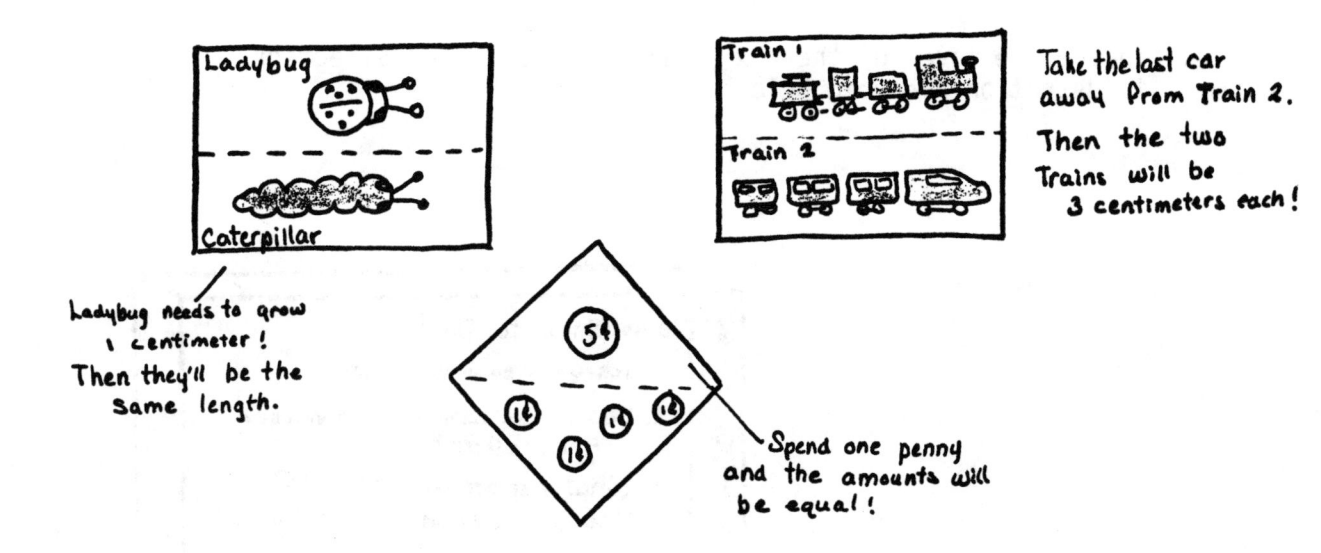

|   |   |   | Equalize |
|---|---|---|---|
| M | 98 Number of Teeth | 6 r 70 number of fingers + toes | Add 22 Pull 3 Teeth |
| N |   |   |   |
| O |   |   |   |

Ladybug needs to grow 1 centimeter! Then they'll be the same length.

Take the last car away from Train 2. Then the two Trains will be 3 centimeters each!

Spend one penny and the amounts will be equal!

44

## PICK A PAK O' PEOPLE

---
Skill: *Completing a sequence of numbers*
---

1. Prepare a box containing several "people paks" such as the one below. (Kids will enjoy helping to prepare the connected cutouts!) Each set should have 7 or more <u>connected</u> cutouts.

2. On the first three or four people in each pak, begin a sequence of whole numbers, fractions, mixed numerals, decimals, etc.

3. Cover the center of each "person" with clear plastic adhesive tape or paper.

4. Inside the cover of the box, place the following directions. See the illustration pictured below:

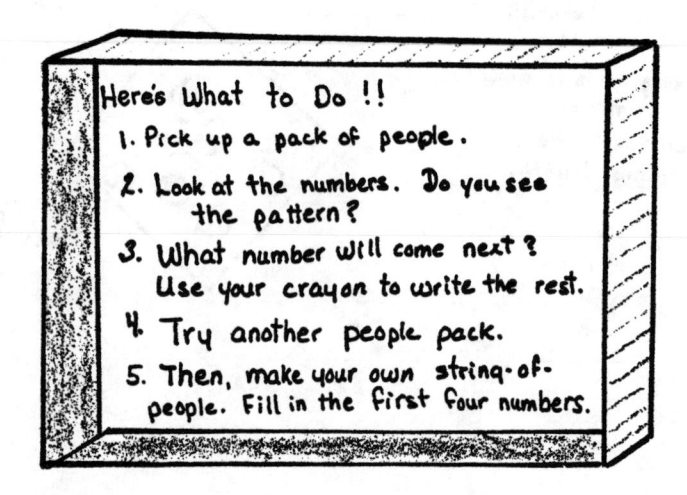

> Here's What to Do !!
> 1. Pick up a pack of people.
> 2. Look at the numbers. Do you see the pattern?
> 3. What number will come next? Use your crayon to write the rest.
> 4. Try another people pack.
> 5. Then, make your own string-of-people. Fill in the first four numbers.

5. Include an answer sheet for self-checking and materials for students to use in making their own people-paks.

STAR TREK

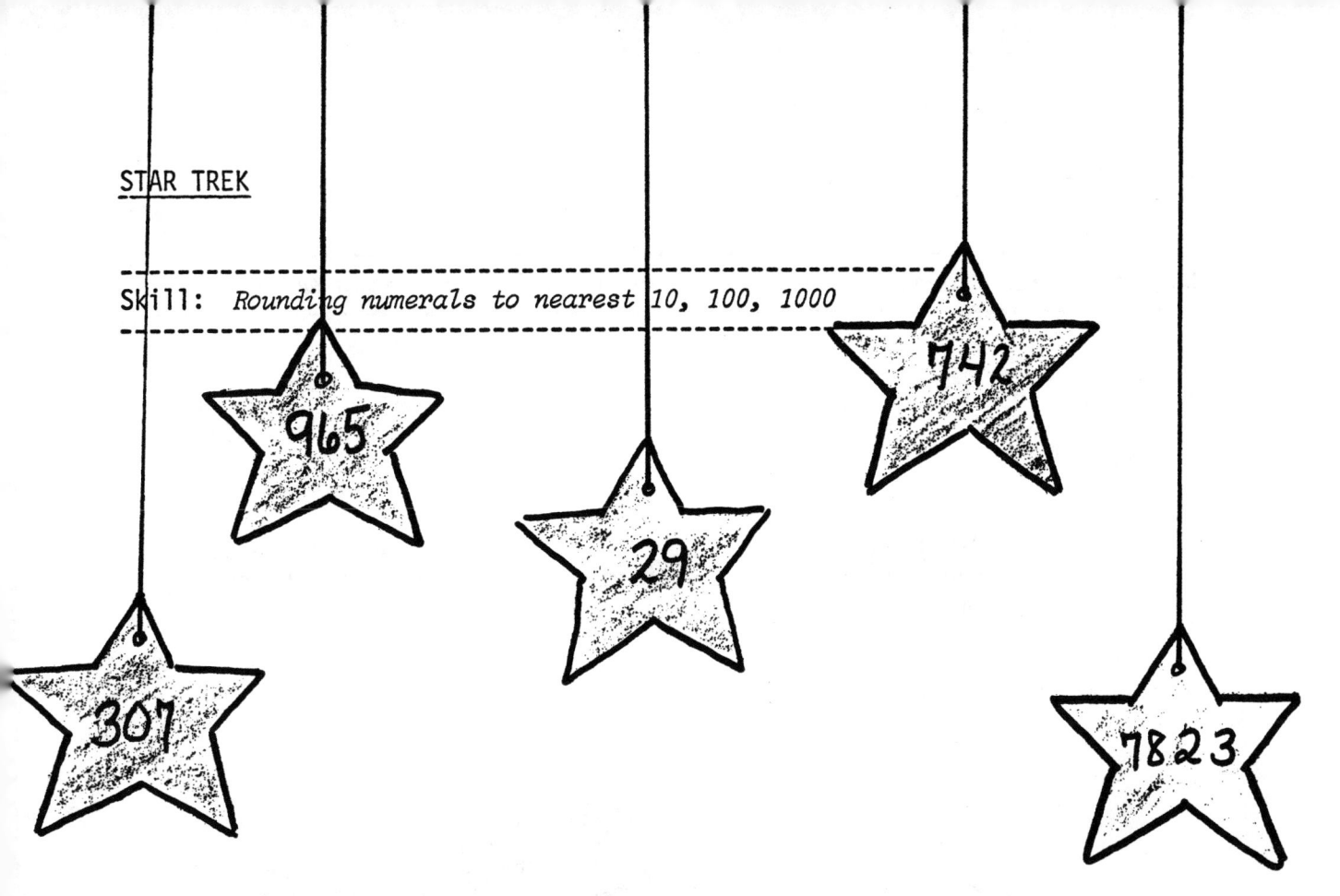

Skill: *Rounding numerals to nearest 10, 100, 1000*

1. All over the classroom, hang numbers from the ceiling (or do this just in your math corner).

2. Duplicate a list of activities such as:

   *Find a number that is close to 400.* _____
   *What number is closer to 760 than 756?* _____
   *Which number is closes to 2710?* _____
   *Round all the numbers to the nearest hundred.*
   *Round all the numbers to the nearest thousand.*

3. Suspend from the ceiling an envelope to hold activity sheets and a CHECK-YOURSELF star.

## DIGIT SWITCH

---
Skill: *Increasing and decreasing values of numbers by given amounts*
---

1. Find a shoe box or a long flat box. Cover the inside of the box top with flannel. Decorate the outside with numbers. Title it DIGIT SWITCH.

2. Cut digits 0-9 from cardboard (several of each). Glue a piece of felt or flannel to the back of each digit. Keep these in the box. Put some commas in the box too.

3. Make a set of cards with directions such as:

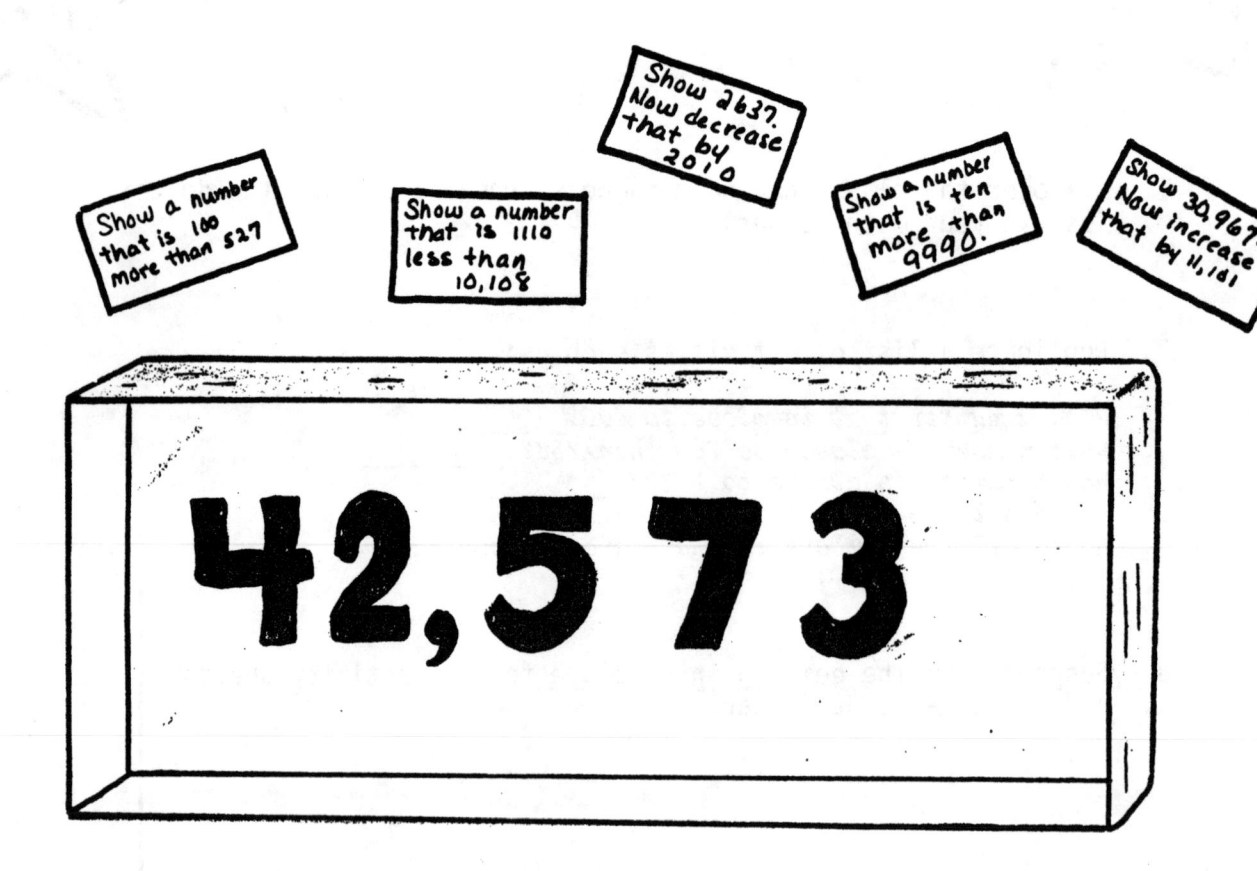

*Show a number that is 100 more than 527*

*Show a number that is 1110 less than 10,108*

*Show 2637. Now decrease that by 2010*

*Show a number that is ten more than 9990*

*Show 30,967. Now increase that by 11,101*

**42,573**

4. The student shows the number on the box-top flannel board, then compares his answer with that on the back of the card.

   Variation: This box will be useful in helping students who need practice just writing numerals or learning place value. Write numbers (in words) and they can practice forming numerals to match. Directions on tape would be suitable for children who cannot yet read the words.

# CLASSROOM ROUND-OFF

---
Skill: *Using a number line to round numerals to nearest 10, 50, 100, 500, 1000, 5000, 10,000, etc.*

---

1. Make a number line that can be taped high on the wall extending all around the room.

2. Number the line by taping numeral labels (counting by 10) above the line. (By not writing on the number line, you'll be able to re-use this again with different kinds of numbers.)

3. Have students draw and cut out objects such as foods, bugs, toys, animals. They should write a number on each picture (not a multiple of 10) and place it along the line at the proper location.

4. Make a series of "rounding" exercises using the objects they've made.

    i.e:  The red 🔍 is closest to what 10?

    Is the △ or the 👞 nearer to 1365?

    What is the location of the 🐭 to the nearest 1,000?

    The ◇ is closest to what hundred?

    Variations:  1. Substitute decimals, positive and negative integers, or fractions for whole numbers simply by putting new labels above the same line.

    2. Use these same materials for estimating missing addends. (i.e. Apple + _____ =caterpillar.)

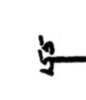

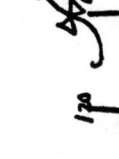

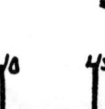

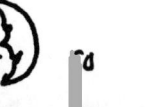

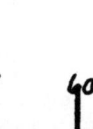

# EXPOSE THE EXPONENT

---
Skill: *Finding the value of numbers written with exponents*
---

1. Make a Matching Board such as the one below. In the first column, list numbers with exponents. In the second column, list their equivalent whole numbers, not in the same order.

2. Students match the exponential numbers in Column A with their values in Column B by placing the peg in the proper hole.

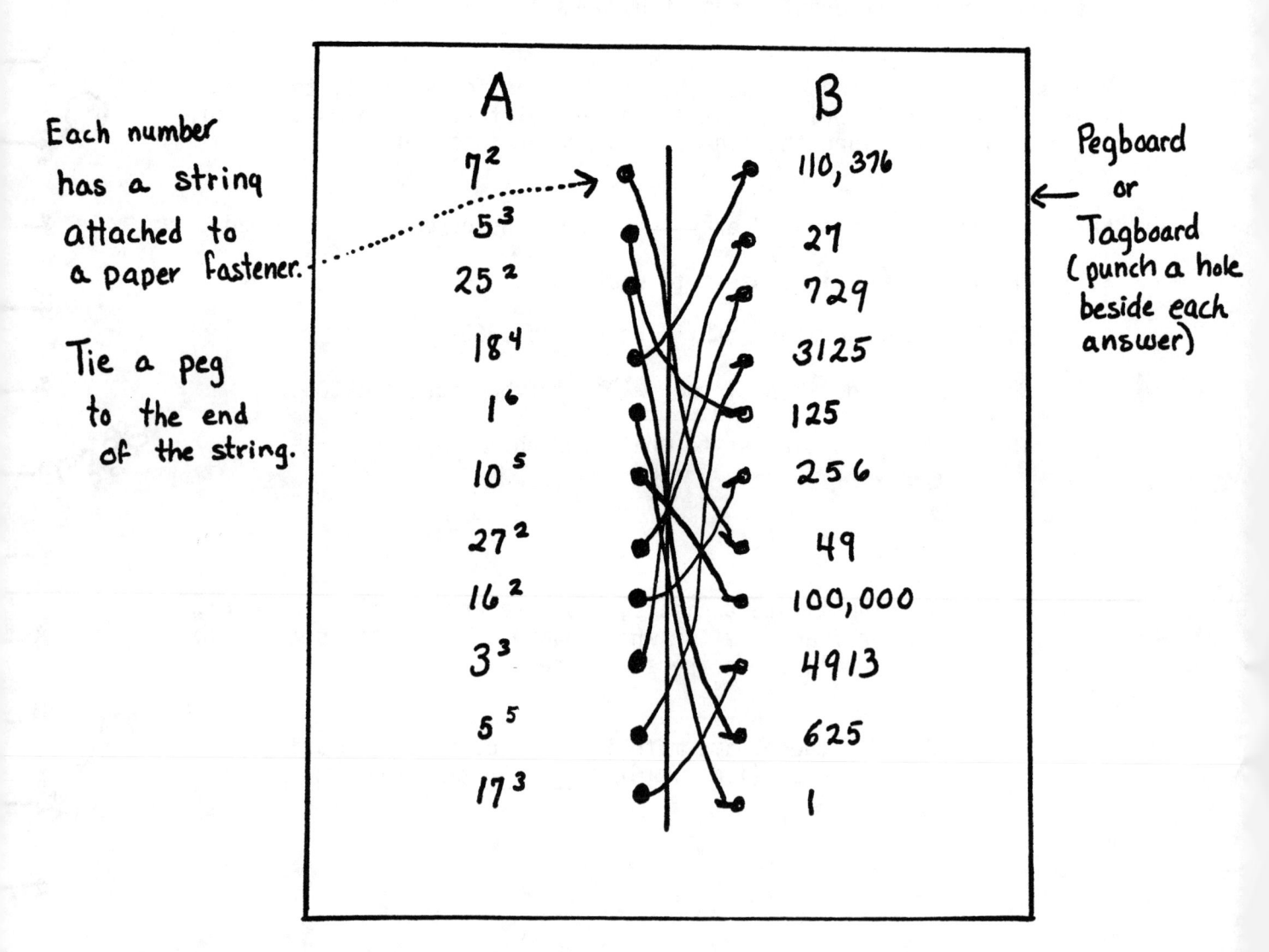

Each number has a string attached to a paper fastener.

Tie a peg to the end of the string.

Pegboard or Tagboard (punch a hole beside each answer)

| A | B |
|---|---|
| $7^2$ | 110,376 |
| $5^3$ | 27 |
| $25^2$ | 729 |
| $18^4$ | 3125 |
| $1^6$ | 125 |
| $10^5$ | 256 |
| $27^2$ | 49 |
| $16^2$ | 100,000 |
| $3^3$ | 4913 |
| $5^5$ | 625 |
| $17^3$ | 1 |

3. Provide a diagram for checking answers.

GALLERY OF ART

---
Skill: *Collecting ideas, words, symbols to represent numbers*
---

1. Supply the following materials:

   paint                       scraps of fabric
   crayons                     glue, scissors
   yarn                        assorted junk
   a big pile of magazines     sandpaper, tissue paper, cardboard

2. Have each student choose a number.

3. Using any of the materials available (plus whatever else he can find) he makes a collage which represents that number.

4. Display these at a "Number Fair" in the classroom or hallway.

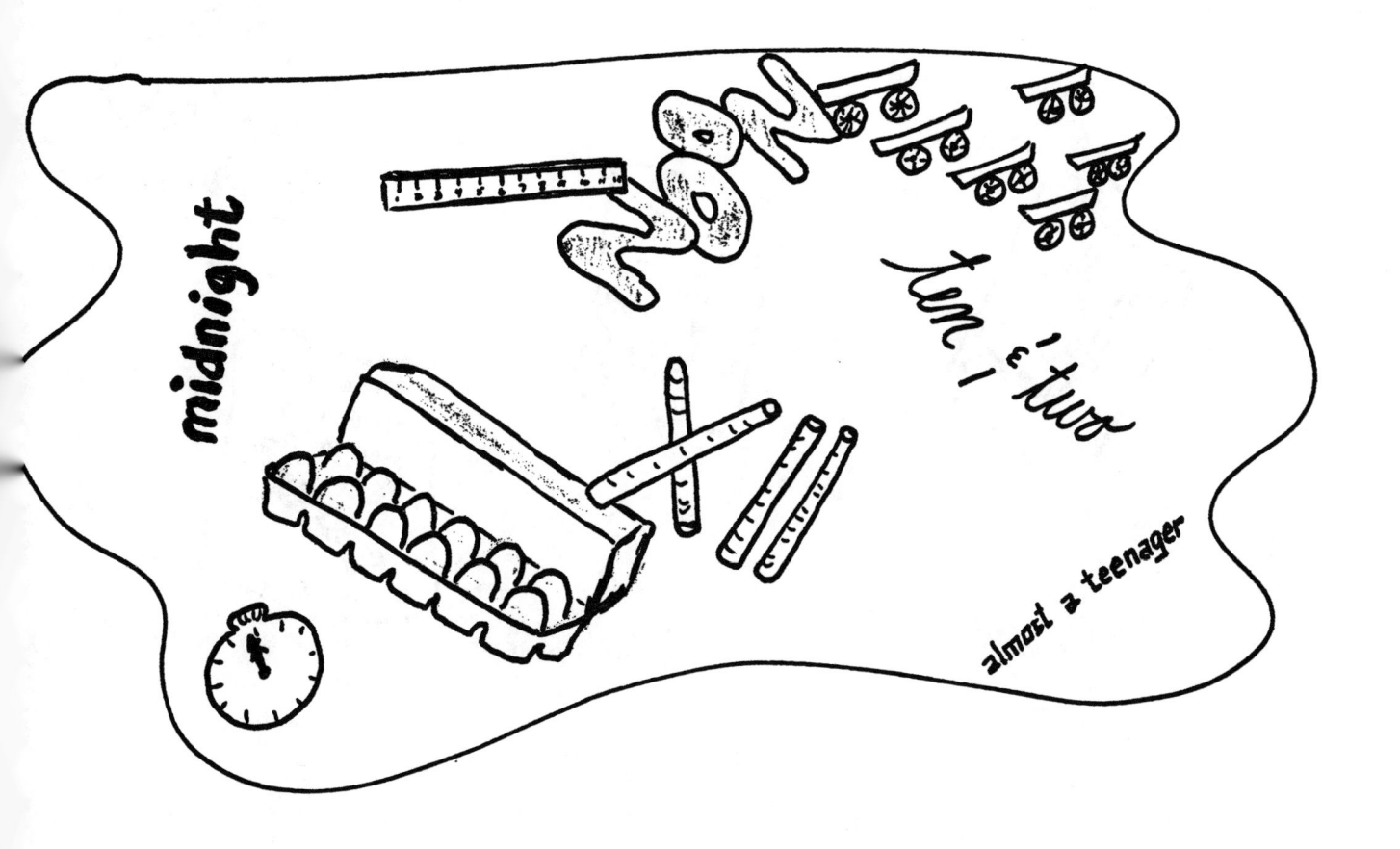

50

## NO NUMBERS HERE!

---
Skill: *Identifying many uses for numbers*
---

1. Ask the class to brainstorm the uses of numbers.

2. Have a secretary keep a list of all the ideas they suggest.

3. Ask students to think about what life would be like without numbers.

4. Each student may take one idea (or more) and make a page showing one way the absence of numbers would make a difference.

5. Put all the pages together to make a story. Add a cover and title and share the book with other classes.

# WHOLE NUMBERS AND INTEGERS

## MATHEMAT

**Skill:** *Learning sums and differences through 20*

1. On a large piece of vinyl (about 5 feet x 6 feet), draw a grid such as the one below. (A shower curtain, plastic tablecloth, or strips of vinyl drawer lining can be written on easily with magic marker.)

2. Create a "mathemat" by writing numbers 0-20 and the symbols +, - and = in the squares.

3. "Hop" a problem such as 13 - 8 = ___. Ask a student to "hop" the answer.

OR

| 7 | 9 | + | 16 | - |
|---|---|---|----|---|
| + | 0 | 5 | 3 | = |
| 15 | = | Use this space for ANY number! | 10 | 13 |
| 20 | 17 | - | 1 | 18 |
| 2 | 6 | 8 | 19 | 12 |

Stand on the answer, and ask student to "hop" a problem!

TRY:  --making problems with 3 or more addends
 --combining addition and subtraction
 --supplying a deck of cards containing all sorts of problems to be "hopped"
 --asking students to make up their own games for two or more players

## ADD A ROW

---
Skill: *Learning basic sums through 20*
---

1. Make several cards like the one below. Each card has four numbers and requires the finding of eight sums.

2. Cover the cards with clear adhesive paper so that students may write on them with crayon.

3. Show the class a sample card and complete the answers together. Tell them to add each row in both directions before they write their answer in the circle.

4. Give each student a card without any numbers. Let him fill in the addends as well as the sums.

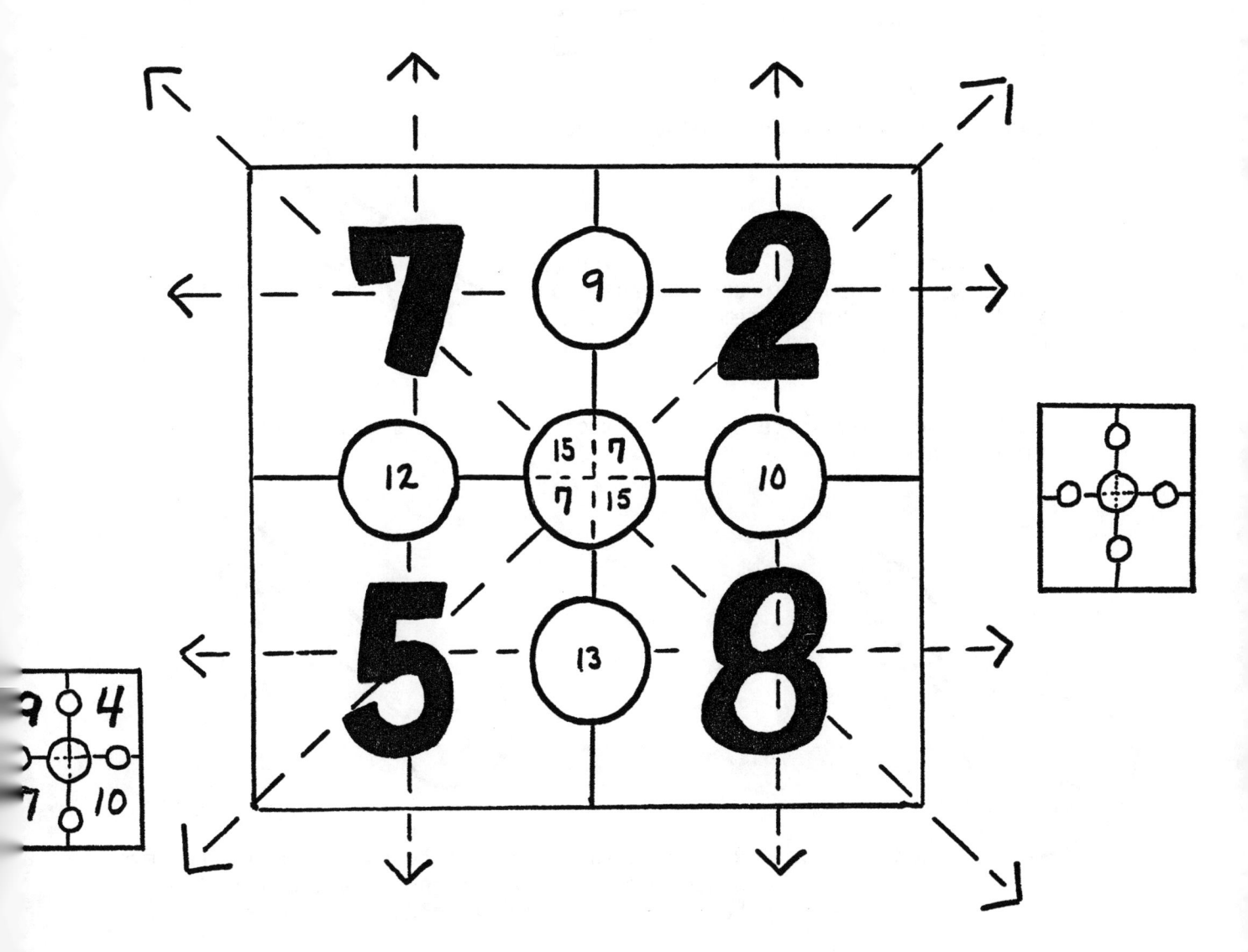

## FINDERS KEY-PERS!

Skill: *Learning basic sums and differences through 20*

1. Cut 21 "keyholes" from tagboard. Label each with the numbers 0-20.

2. Pin the "keyholes" on a piece of corkboard that a student could carry to his own working place.

3. Make several keys, each bearing an addition or subtraction problem having an answer from 0 through 20.

4. Place these keys in an envelope pinned to the board. Include some blank keys.

5. The student finds all the keys which could "unlock" the answer in each keyhole, and pins those keys next to it (or hangs them on a hook above it).

6. Each student may write one problem of his own on a blank key to add to the envelope.

Alternative: Reproduce sheets of the keyholes and of the keys. Students might cut out the keys and glue them beneath the proper keyhole.

55

MR. NOBODY

---
Skill: *Recognizing zero as the identity element for addition*
---

*(speech bubbles: "I'm some fun, but I don't change the sum." "I'm Nothing!")*

1. Give each student a number square such as the one below. (Use any numbers, but be sure to put 0 in the center and in each row.)

2. For each horizontal, vertical, and diagonal row of numbers, students write and solve an addition equation.

3. At the bottom of each card, provide a place for the student to write a sentence describing the function of zero in addition.

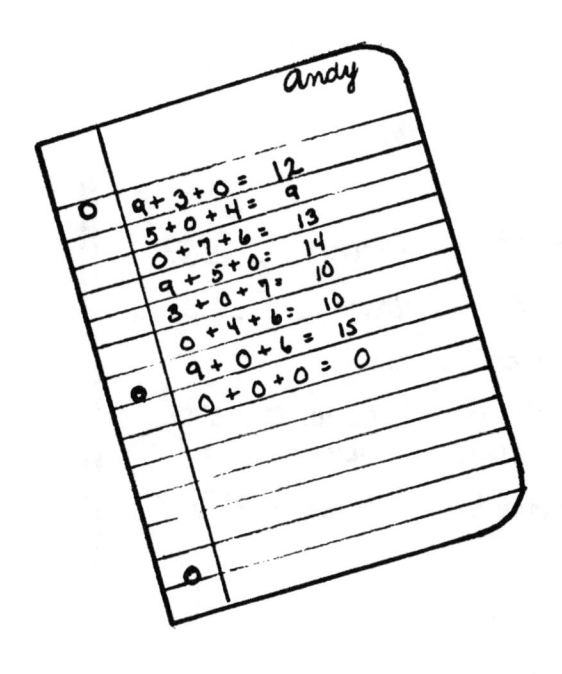

Andy

9 + 3 + 0 = 12
5 + 0 + 4 = 9
0 + 7 + 6 = 13
9 + 5 + 0 = 14
3 + 0 + 7 = 10
0 + 4 + 6 = 10
9 + 0 + 6 = 15
0 + 0 + 0 = 0

| 9 | 3 | 0 |
| 5 | 0 | 4 |
| 0 | 7 | 6 |

Tell how zero acts in addition problems: ↓

56

## WORM WISDOM

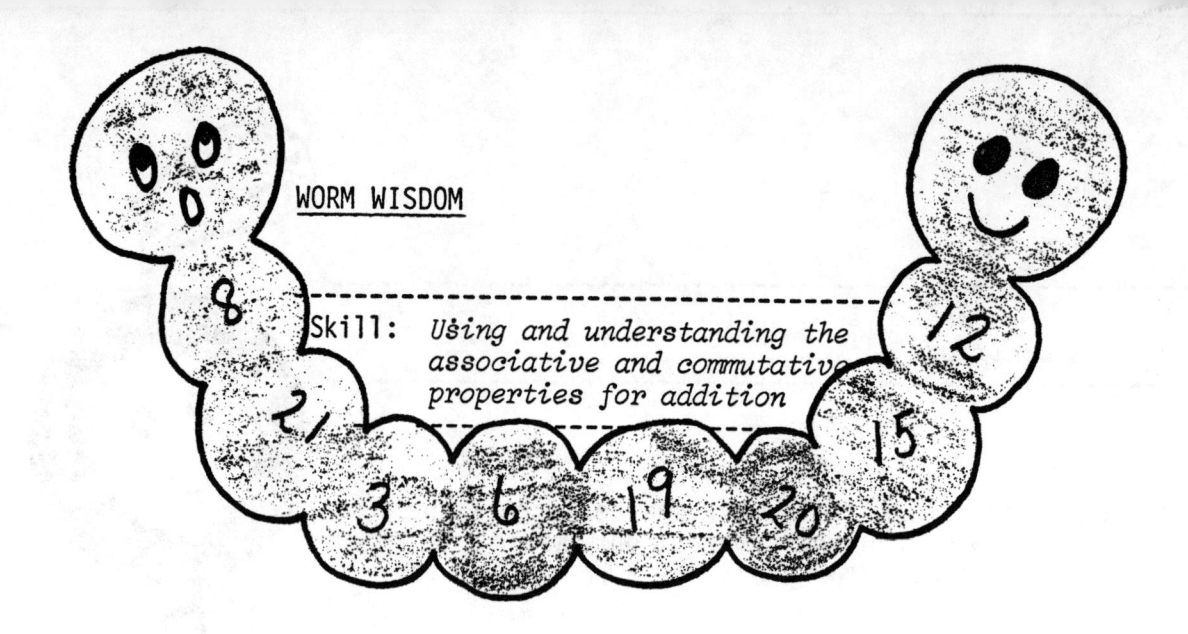

**Skill:** *Using and understanding the associative and commutative properties for addition*

(Worm sections show: 8, 2, 3, 6, 19, 20, 15, 12)

1. Make a long two-headed worm or other creature. Write an addend on each section. Put a string on him so that he can be carried to a private place.

2. Give students the following directions (see illustration below).

   Variations:
   1. Use larger addends for older students.
   2. Make enough worms for each student to have one, or let them make their own.
   3. Give identical worms to two students or teams and let them race to see who can find the sum first.

---

Name _____

Add from left to right.   Answer _____.
Add from right to left.   Answer _____.
Does it matter which way you add?
How many addition problems can you write using the addends on the worm? Write some.
What is the fastest way to add these addends?

# CROCK-A-CODE

**Skill:** *Finding missing addends and parts of addends in addition equations*

Write a special message to your class using numbers to represent letters. As each missing addend is found, students discover which letter should fill the blank.

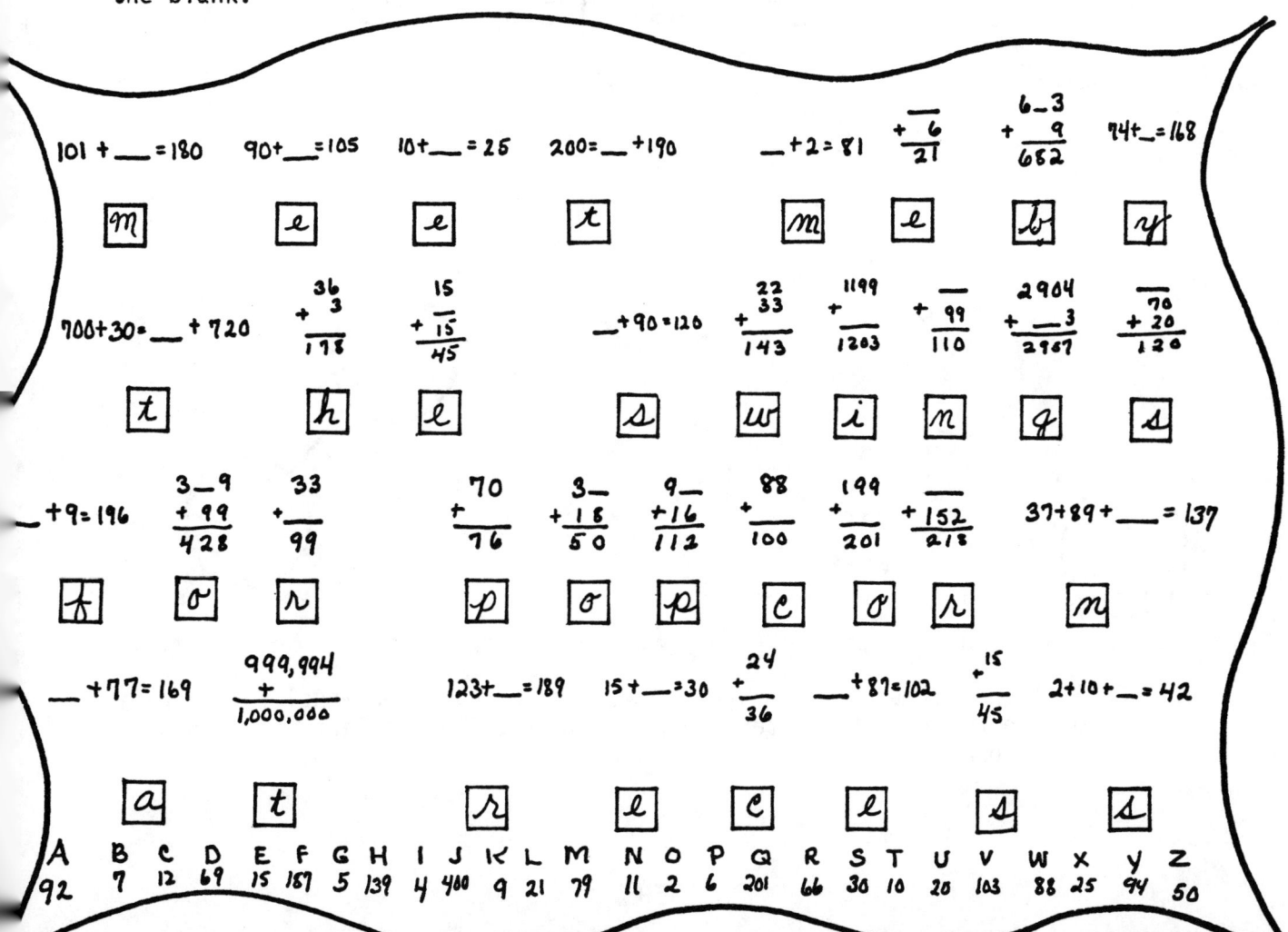

$101 + \_\_ = 180$   $90 + \_\_ = 105$   $10 + \_\_ = 25$   $200 = \_\_ + 190$   $\_\_ + 2 = 81$   $\begin{array}{r}\overline{\phantom{0}6} \\ + 6 \\ \hline 21\end{array}$   $\begin{array}{r}6\_3 \\ 9 \\ +\_\_ \\ \hline 682\end{array}$   $94 + \_\_ = 168$

[m]  [e]  [e]  [t]     [m]  [e]  [b]  [y]

$700 + 30 = \_\_ + 720$   $\begin{array}{r}36 \\ 3 \\ +\_\_ \\ \hline 178\end{array}$   $\begin{array}{r}15 \\ +\overline{15} \\ \hline 45\end{array}$   $\_\_ + 90 = 120$   $\begin{array}{r}22 \\ 33 \\ +\_\_ \\ \hline 143\end{array}$   $\begin{array}{r}1199 \\ +\_\_ \\ \hline 1203\end{array}$   $\begin{array}{r}\overline{\phantom{0}} \\ +99 \\ \hline 110\end{array}$   $\begin{array}{r}2904 \\ +\_\_3 \\ \hline 2907\end{array}$   $\begin{array}{r}\overline{70} \\ +20 \\ \hline 120\end{array}$

[t]  [h]  [e]   [s]  [w]  [i]  [n]  [g]  [s]

$\_\_ + 9 = 196$   $\begin{array}{r}3\_9 \\ +99 \\ \hline 428\end{array}$   $\begin{array}{r}33 \\ +\_\_ \\ \hline 99\end{array}$   $\begin{array}{r}70 \\ +\_\_ \\ \hline 76\end{array}$   $\begin{array}{r}3\_ \\ +18 \\ \hline 50\end{array}$   $\begin{array}{r}9\_ \\ +16 \\ \hline 112\end{array}$   $\begin{array}{r}88 \\ +\_\_ \\ \hline 100\end{array}$   $\begin{array}{r}199 \\ +\_\_ \\ \hline 201\end{array}$   $\begin{array}{r}\overline{\phantom{0}} \\ +152 \\ \hline 218\end{array}$   $37 + 89 + \_\_ = 137$

[f]  [o]  [r]   [p]  [o]  [p]  [c]  [o]  [r]  [n]

$\_\_ + 77 = 169$   $\begin{array}{r}999,994 \\ +\_\_ \\ \hline 1,000,000\end{array}$     $123 + \_\_ = 189$   $15 + \_\_ = 30$   $\begin{array}{r}24 \\ +\_\_ \\ \hline 36\end{array}$   $\_\_ + 87 = 102$   $\begin{array}{r}15 \\ +\_\_ \\ \hline 45\end{array}$   $2 + 10 + \_\_ = 42$

[a]  [t]     [r]  [e]  [c]  [e]  [s]  [s]

| A | B | C | D | E | F | G | H | I | J | K | L | M | N | O | P | Q | R | S | T | U | V | W | X | Y | Z |
|---|---|---|---|---|---|---|---|---|---|---|---|---|---|---|---|---|---|---|---|---|---|---|---|---|---|
|92 | 7 |12 |69 |15 |187| 5 |139| 4 |400| 9 |21 |79 |11 | 2 | 6 |201|66 |30 |10 |20 |103|88 |25 |94 |50 |

ADD AND EAT!

---

Skill: *Solving equations with several addends*

---

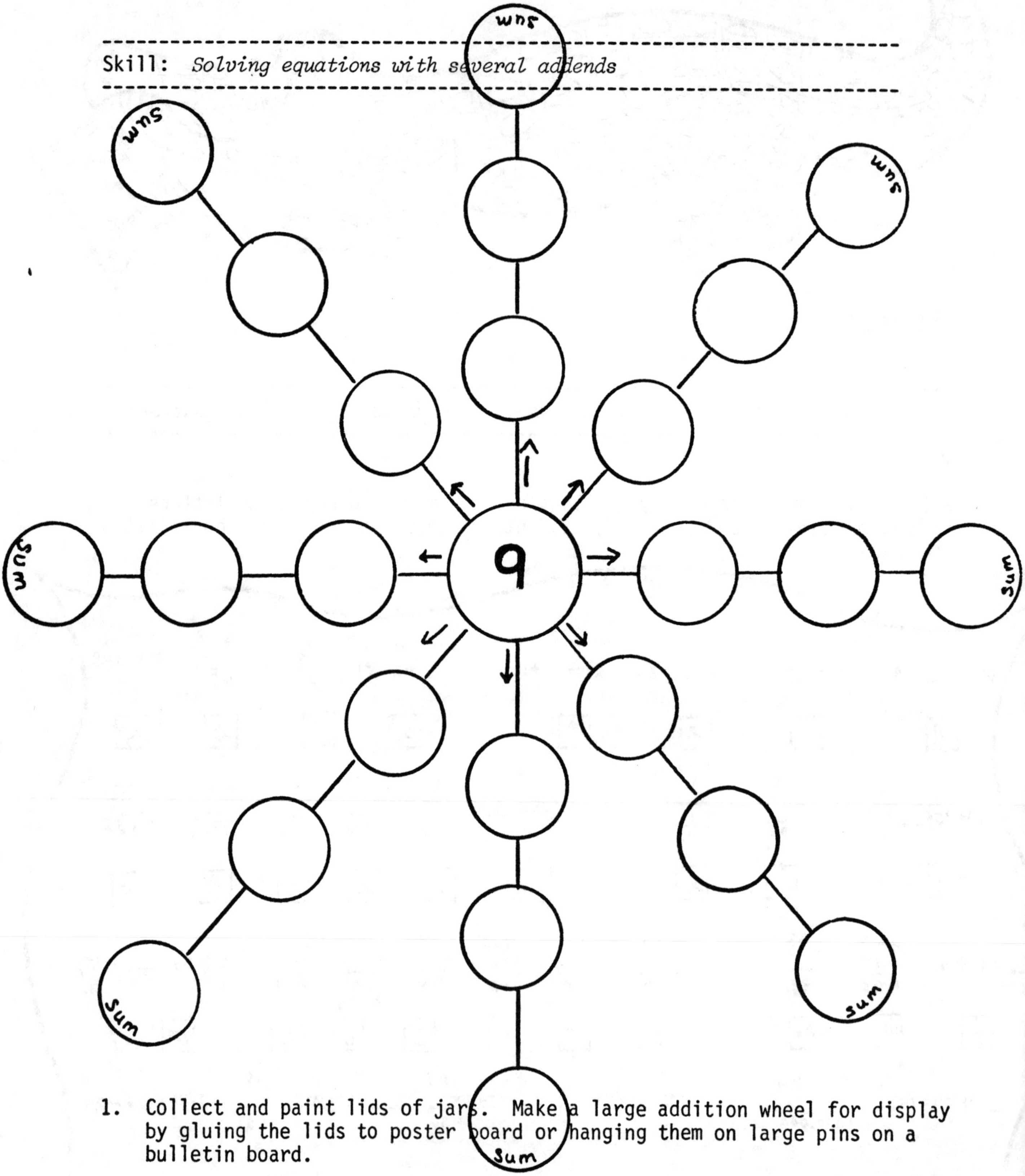


1. Collect and paint lids of jars. Make a large addition wheel for display by gluing the lids to poster board or hanging them on large pins on a bulletin board.

2. Write a number on the center lid and on each of the two lids closest to the center.

3. Ask students to find the sum of each spoke by adding out from the center.

4. You may wish to ask students to write an equation for each spoke also.

   Variations: 1. Have students make their own addition wheels by gluing pipe cleaners to bottle caps, old plastic discs, or anything of a circular shape.

   2. Ask them to try addition circles on which each spoke has the same sum as the others.

   3. Try "Add and Eat Wheels." Duplicate a sheet with a blank addition wheel. On the blanks, students arrange lifesavers. Each color may stand for a certain number. They write the number in the center of the lifesaver, add the addends, then rearrange to make new problems.

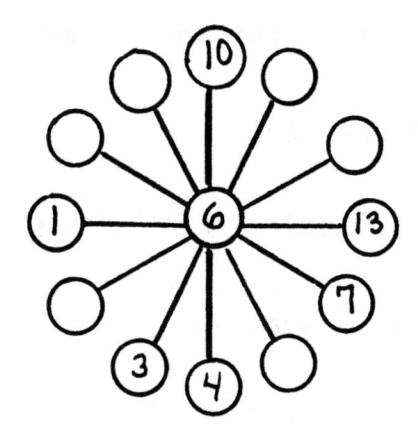

   When the math exercise is finished, the mathematicians may eat the wheels!

# GASTRONOMIC GUESSTIMATION

**Skill:** *Estimating sums and differences*

1. Fill several jars to different levels with jellybeans, marshmallows, gumdrops, etc.

2. Have students estimate the number in each jar.

3. Count the contents of the jars to check the accuracy of the estimates.

4. Label each jar with the actual number.

5. Then give students a list of problems such as these:

    Estimate the following sums and differences:

    A + B
    C + E
    A + D + F
    A + B + C + D + E + F
    D - C
    (E + A) - F
    B + C + (D - A)

6. When everyone is finished? ? ? Eat the evidence, of course!

## SUM STAIRS!

---
Skill: *Adding and subtracting several numbers in one equation*
---

1. Ask students to cut out stairsteps.

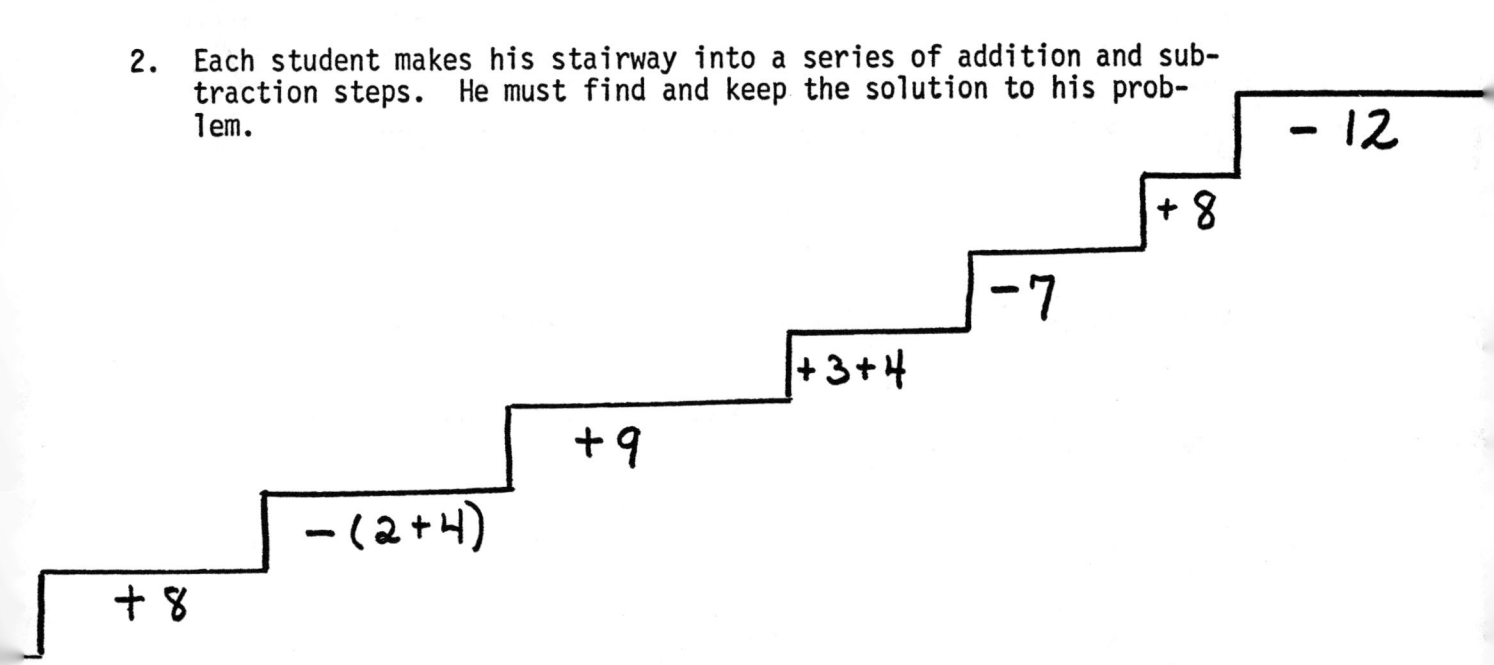

2. Each student makes his stairway into a series of addition and subtraction steps. He must find and keep the solution to his problem.

```
                                                    ┌─ -12
                                              ┌─ +8 ┘
                                        ┌─ -7 ┘
                                 ┌─+3+4 ┘
                          ┌─ +9 ┘
                ┌─ -(2+4)┘
          ┌─ +8 ┘
```

3. Students trade stairs and each writes an equation to solve the one he has received.

4. The "creator" of each stairway checks the answer of anyone who solves his problem.

5. Students may continue trading and solving. You might offer a prize to anyone who climbs all the stairs.

   Variation: Stairways can be more challenging if the answers are provided, but some steps left blank!

SEE-SAW SUMS

---
Skill: *Using different addends to obtain the same sum*
---

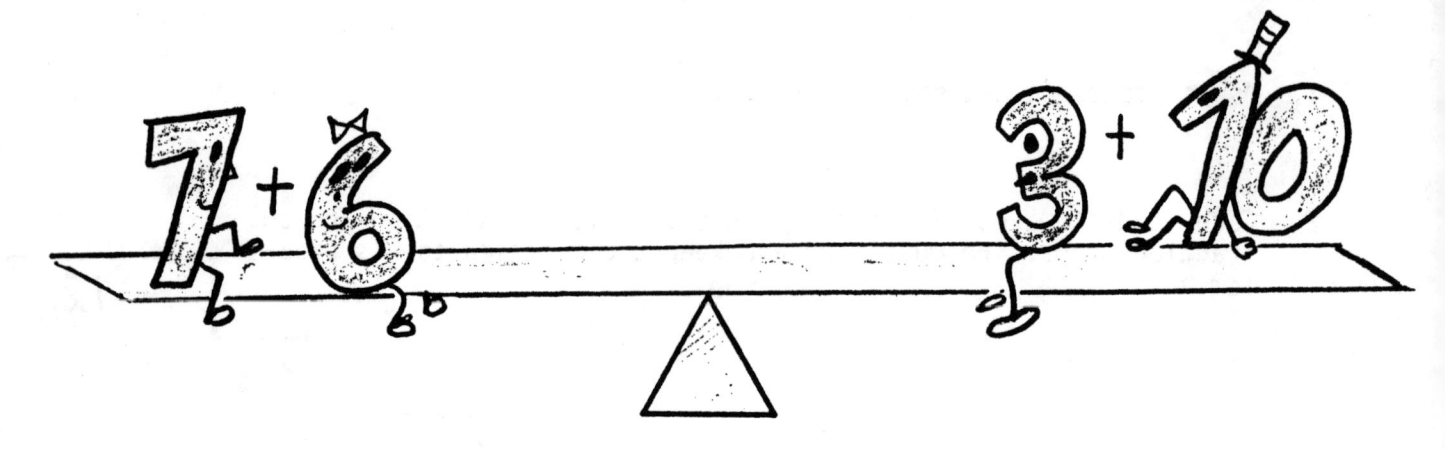

*7 + 6*     *3 + 10*

1. On a bulletin board, display a see-saw or a balance beam.

2. Write addition problems on little people and pin them around the edges of the board.

3. In a pocket, place a problem sheet with questions such as are illustrated below.

4. As students find a pair of people that will balance each other, they pin them on the see-saw.

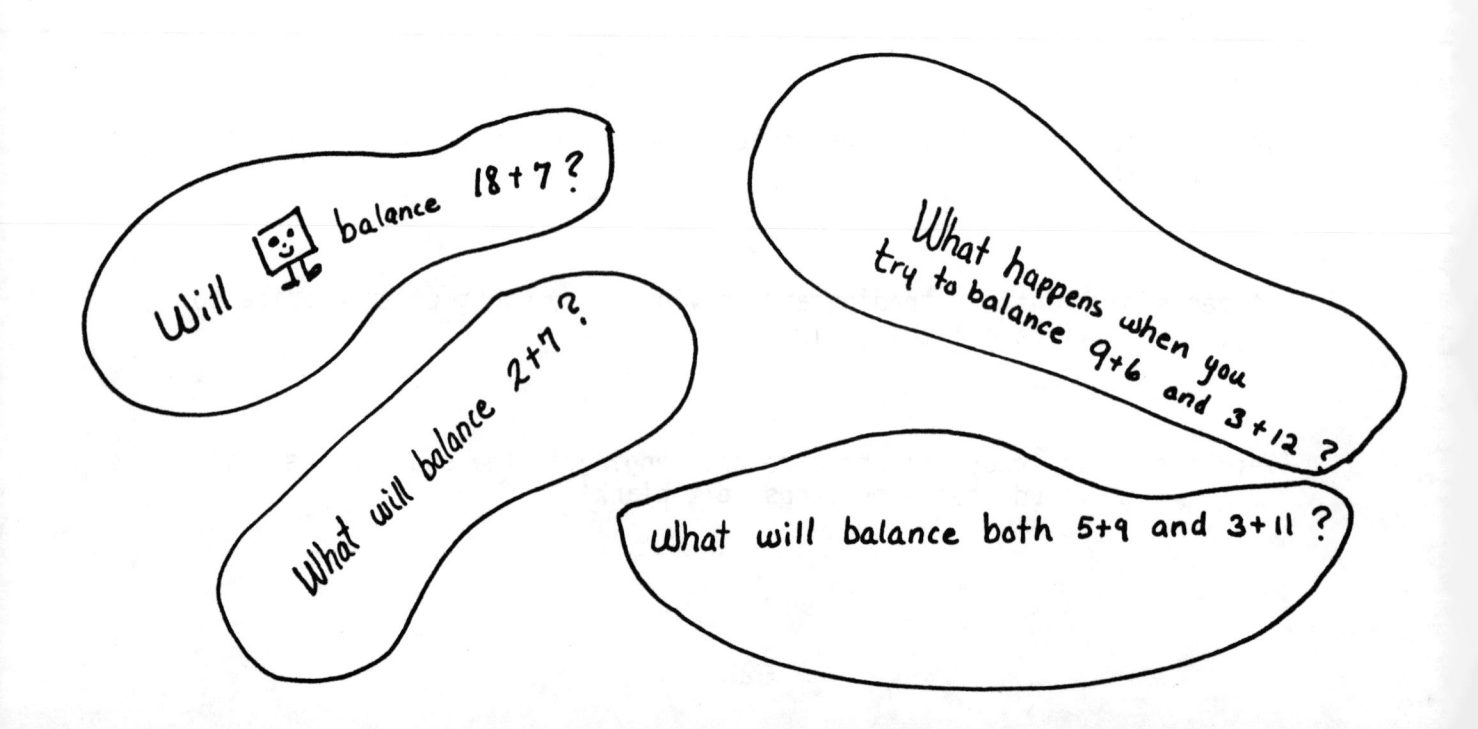

Will [☐] balance 18 + 7?

What will balance 2 + 7?

What happens when you try to balance 9 + 6 and 3 + 12?

What will balance both 5 + 9 and 3 + 11?

# PALINDROME PARTY

**Skill:** *Recognizing and using patterns in addition*

Give students these directions:

> Write 428.
> Now add to it the number backward.
> Now write the sum backward and add again.
> What is special about this sum?
>
> Now try 67.
> Turn it around and add.
> Turn the sum around. Add again.
> What does the sum look like?

428
824
252
521
773

67
+76
143
+341
484

Numbers that read the same forward or backward are called PALINDROMES.

You can find palindromes by adding together reversed numbers.

Sometimes you have to add several times before you find a palindrome. Remember to reverse the number each time before you add.

4397

**Variation:** Explain the concept of palindromic numbers in a learning center. Show this process for finding palindromes and give the students several numbers with which to work.

Try it with 2184

## SUBTRACTION 500

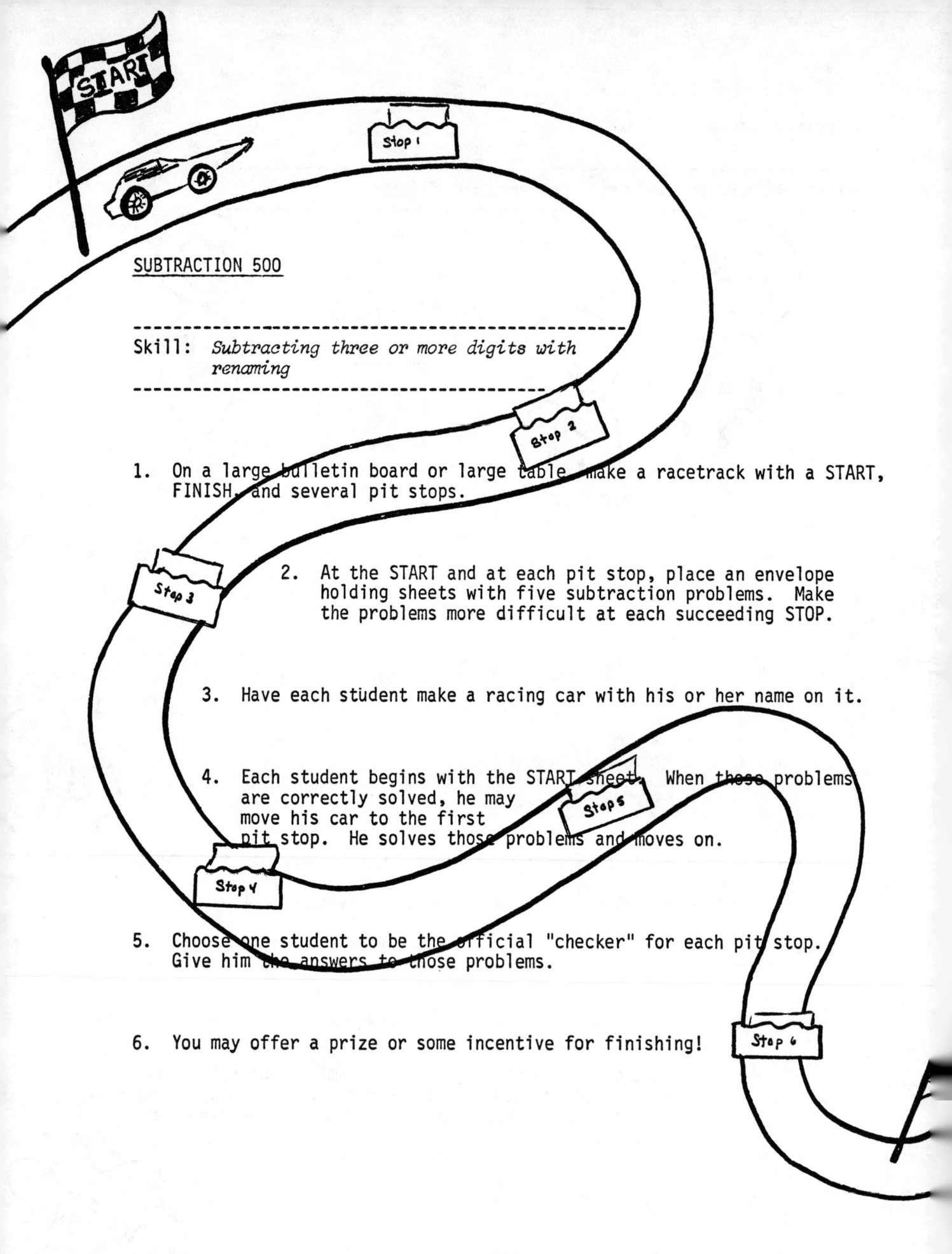

---
Skill: *Subtracting three or more digits with renaming*
---

1. On a large bulletin board or large table, make a racetrack with a START, FINISH, and several pit stops.

2. At the START and at each pit stop, place an envelope holding sheets with five subtraction problems. Make the problems more difficult at each succeeding STOP.

3. Have each student make a racing car with his or her name on it.

4. Each student begins with the START sheet. When these problems are correctly solved, he may move his car to the first pit stop. He solves those problems and moves on.

5. Choose one student to be the official "checker" for each pit stop. Give him the answers to those problems.

6. You may offer a prize or some incentive for finishing!

# TATTLE-TAPES

**Skill:** *Adding long columns of numbers*

1. Have students collect cash register tapes from grocery store shopping trips.

2. Cover or remove the totals on the tapes. Let students be the "adding machines" and find the totals. The totals will "tattle" on their machines by revealing the speed and accuracy of their addition skill.

3. Begin with shorter tapes and gradually add longer ones.

4. Try covering some of the numbers within the columns instead of covering the totals. Then ask students to find what digits might be missing in the problems.

Variation: Make a file of cash register tapes covered with plastic. Arrange them in order of difficulty, including those with missing addends near the end. Students can fill in numbers, "wipe off" their answers, and try another one!

## BATTER UP!

**Skill:** *Renaming tens, hundreds, and thousands*

1. Make baseballs from tagboard. On each ball write a 2, 3, 4, or 5-digit numeral.

2. Make four bats from tagboard also. Write these directions on the bats:

    a. Rename as ones.
    b. Rename as tens.
    c. Rename as hundreds.
    d. Rename as thousands.

3. Split students into teams, decide on the location of bases, and begin the game as follows:

    The pitcher "pitches" (reads) a ball to the batter. The batter uses whichever bat he chooses (or whichever one he is asked to try). He must follow the directions on his bat. If he does the renaming correctly, he moves:

    --one base for renaming as ones or tens

    --two bases for renaming as hundreds or thousands

    He may get a HOME RUN for correctly renaming numbers including zeros (i.e., 4700 = 4690 + 10 ones)

# BLOCKBUSTER

---
Skill: *Subtracting with zeros in the minuend*
---

1. Find a large square box. Decorate it with brightly colored fabric, paper, yarn, etc.

2. Write or paste ten (or more) numbers with zeros on each of five of the box's six sides.

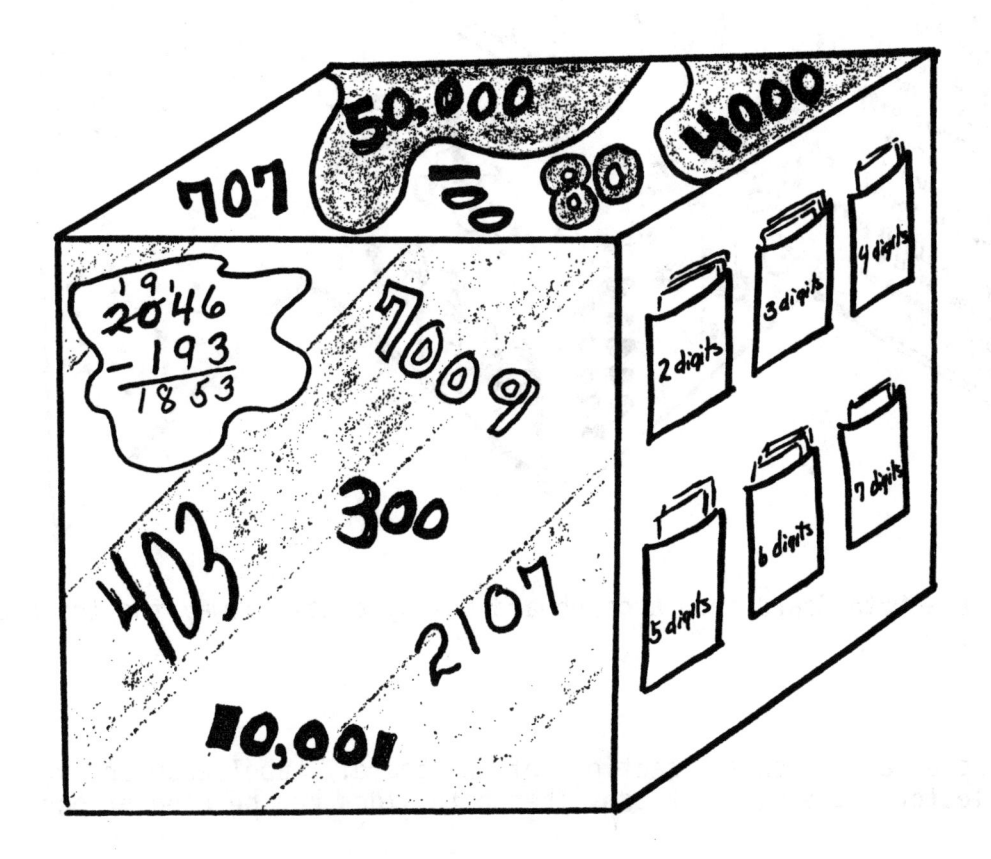

3. On the remaining side of the box, attach six pockets. In the pockets, place numbers to subtract.

4. On the box, attach examples of problems showing how to do renaming with zeros. Show several different kinds of problems.

5. Direct students to choose a number (a minuend) from a side of the box, then to take a number (subtrahend) from one of the pockets. For each pair of numbers, they write and solve a subtraction problem using the method shown on the box.

# WAYS WITH ARRAYS

**Skill:** *Writing and solving related addition and subtraction equations from arrays*

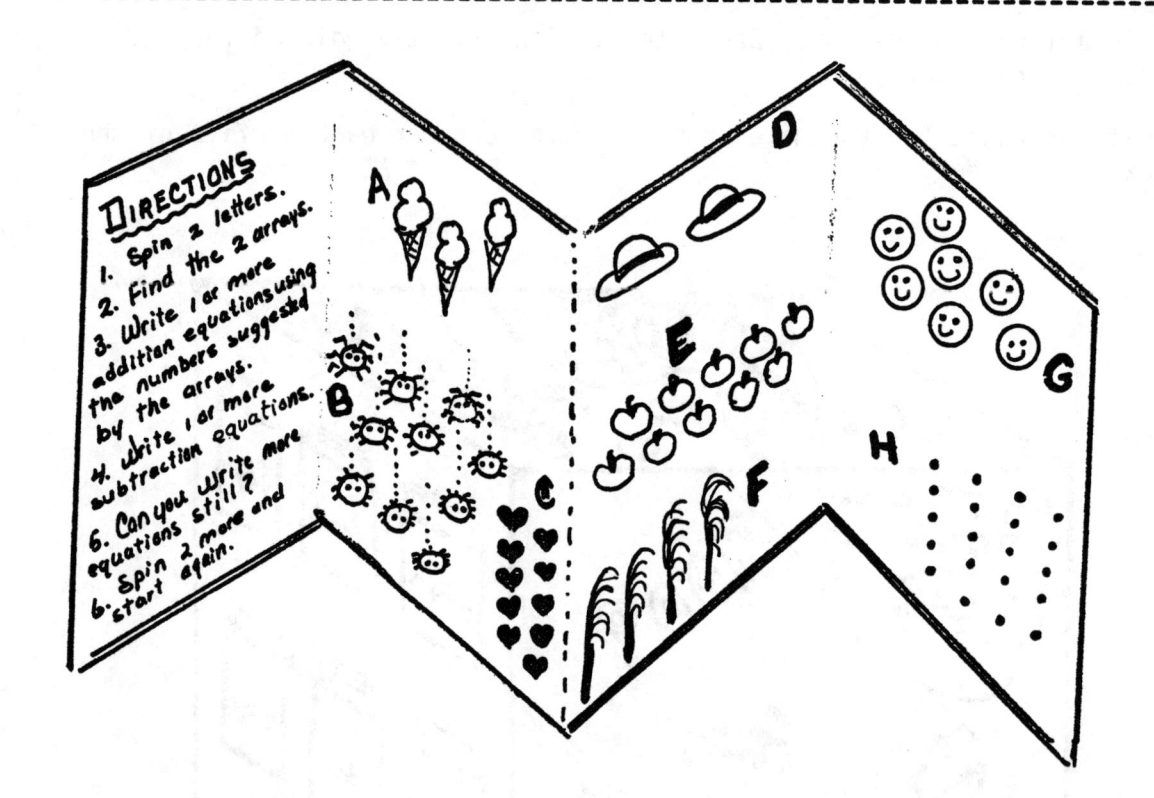

*(Illustration of a folded cardboard Math Board with arrays labeled A–H: ice cream cones, bugs, hearts, hats, apples, wheat, smiley faces, dots. Directions on the board read:)*

**DIRECTIONS**
1. Spin 2 letters.
2. Find the 2 arrays.
3. Write 1 or more addition equations using the numbers suggested by the arrays.
4. Write 1 or more subtraction equations.
5. Can you write more equations still?
6. Spin 2 more and start again.

1. Make a Math Board from a cardboard box by cutting away the top and bottom.

2. Display arrays on all sections of the board. Label each array with a letter. Use the back too, then both sides can be used at once!

3. Put the letters on a spinner. Make one spinner for each side.

4. Attach the above directions to the front and back.

Variation: Spinning three or more letters will challenge students to write more complicated equations!

Example:

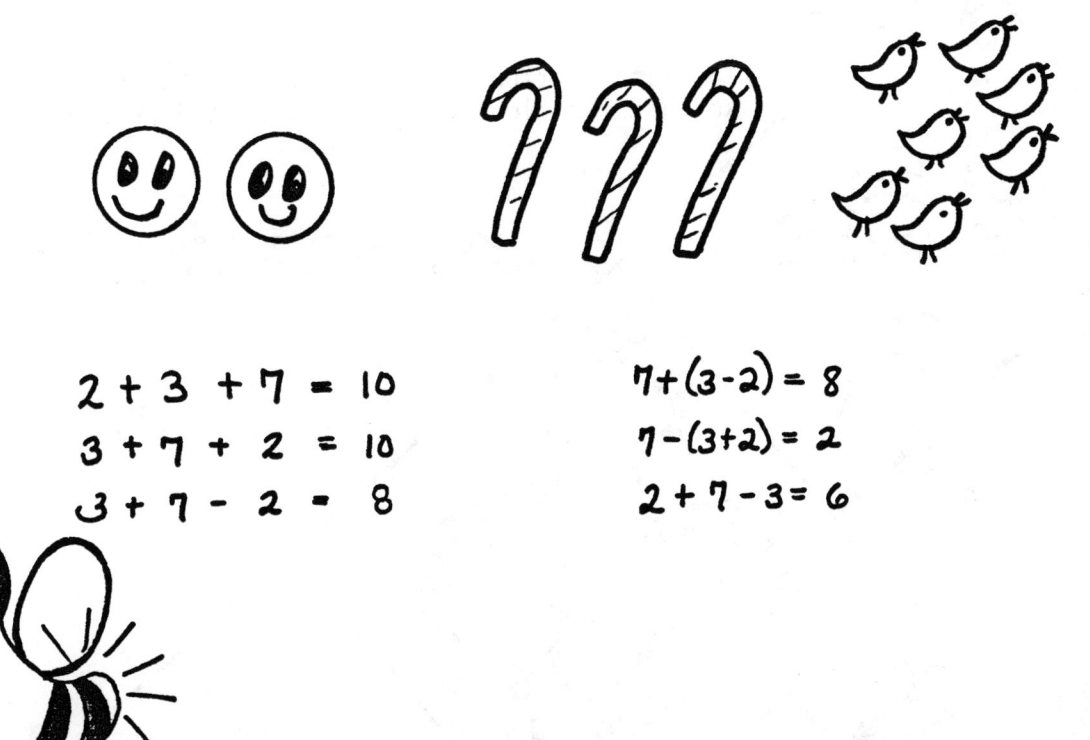

$2 + 3 + 7 = 10$
$3 + 7 + 2 = 10$
$3 + 7 - 2 = 8$

$7 + (3-2) = 8$
$7 - (3+2) = 2$
$2 + 7 - 3 = 6$

Sum Tic-Tac-Toe: Start with a numbered grid (digits 1-9). Each player selects a sum (6-24) and writes it in a secret place. He tries to place his X's (or O's) on three addends that will yield that sum. (The marks need not be in rows.) When the sum is reached, the game is over!

| 6 | 3 | ⓧ |
|---|---|---|
| 1 | 4 | ✗ |
| 7 | ✗ | 8 |

# DRILLSTRIPS

---
Skill: *Writing and solving word problems requiring use of addition and subtraction*
---

1. Give each student a piece of adding machine tape about a foot long.

2. Ask each student to write (and represent with illustrations) an addition or subtraction problem. They may also try problems with more than one step.

3. The last section could be covered with clear adhesive so that equations can be written and erased.

4. Students may trade their Drillstrips and solve each other's problems.

**bright idea!**

Try this!  
1 x 9 + 2 = 11  
12 x 9 + 3 = 111  
123 x 9 + 4 = 1111  
1234 x 9 + 5 =

Finish the pattern!

*(Drillstrip illustration: "This is Tom's mother. She found 4 ants in her cupboard. Dan's mom discovered ants in her bread box. Who had more ants? How many more?")*

71

IT'S A MOD, MOD WORLD

---
Skill: *Adding and subtracting in a modular number system*
---

1. Use a large circular "pizza cardboard" to make a Mod 5 clock. Cut a hand from cardboard and attach it to the clock with a paper fastener.

2. Introduce this to the class by asking questions such as:

   *Start at 0. Move five spaces. Where do you stop?*
   *What addition equation could you write for the problem you just did?*

   *Start at 0. Move seven spaces. Where do you stop?*
   *0 + 7 = ___?___ in Mod 5.*

   *Use the clock to add 3 + 4.*

   *How would you use the clock to subtract 4 from 6?*
   *(Move six spaces from zero, then go backward 4.)*

   *How would you make a Mod 7 clock?*

72

3. Have each student make his own Mod clock, using any one-digit number above 1. Paper plates or small "pizza circles" make good individual clocks.

4. Supply a list of addition and subtraction problems to be solved using the clocks they've made.

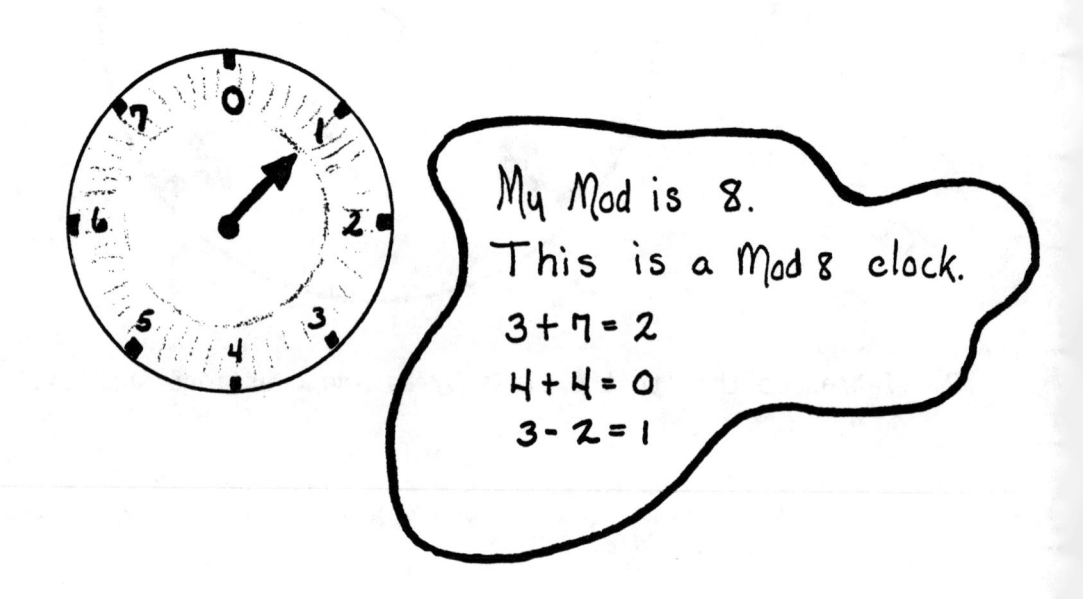

My Mod is 8.
This is a Mod 8 clock.
3 + 7 = 2
4 + 4 = 0
3 - 2 = 1

NOTE: The glossary will aid you in explaining Mods to students.

## GRID SPECIAL

**Skill:** *Visualizing multiplication as combining subsets into a set*
*Visualizing division as separating sets into subsets*

$2 \times 4 = 8$

1. Make a grid on an overhead transparency. Give graph paper to the students.

2. Work together to picture some multiplication problems on the graph paper.

3. Ask students to discover how they could show division. Try some examples like the one pictured at right.

4. Give a number of other multiplication and division problems to be shown on graph paper. Direct students to show an equation for each picture.

   * Try some division equations with remainders!

$12 \div 3 = 4$

## WIN-A-PIN!

---
Skill: *Learning basic multiplication and division facts using factors through 9*
---

1. Prepare the materials for playing Win-a-Pin.

    --Find a long, narrow box (2 ft. x 1 ft.). Along one edge hang a heavy cord. Attach it at the ends with paper fasteners or heavy tape.

    --Cover the inside bottom of the box with felt.

    --Write products from 0-81 on clothespins.

    --On two small wooden blocks, write the numbers 0-5.

    --Provide a chart with all the facts for checking answers.

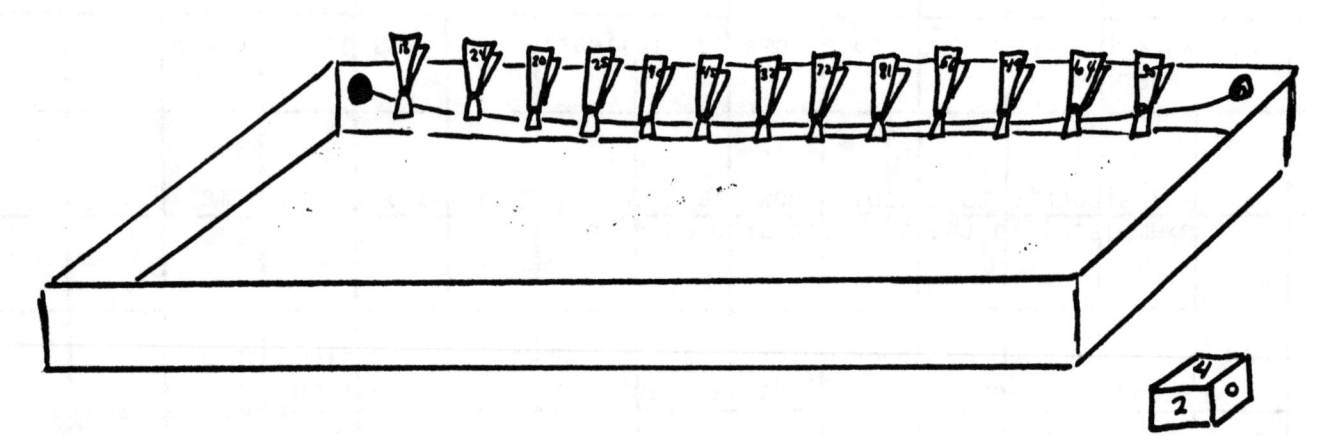

2. Students clip the clothespins along the card.

3. Taking turns, each player rolls both cubes twice and finds the clothespin that shows the product of those two rolls. He keeps that clothespin if his answer is correct. If a product has been used, the player tosses again.

4. Play continues until all the clothespins are taken. The winner is the person who has collected the most clothespins.

    Variation: If students need practice just with 1-5, give them just two cubes and the products 0-25. They roll each cube once to find their factors.

HOT TO DOT!

---
Skill: *Supplying missing factors and products in multiplication and division equations*
---

Duplicate a dot-to-dot picture such as the one below and a set of directions for each student. Arrange the equations in such a way that a line drawn from A-B-C-D, etc. will complete the picture.

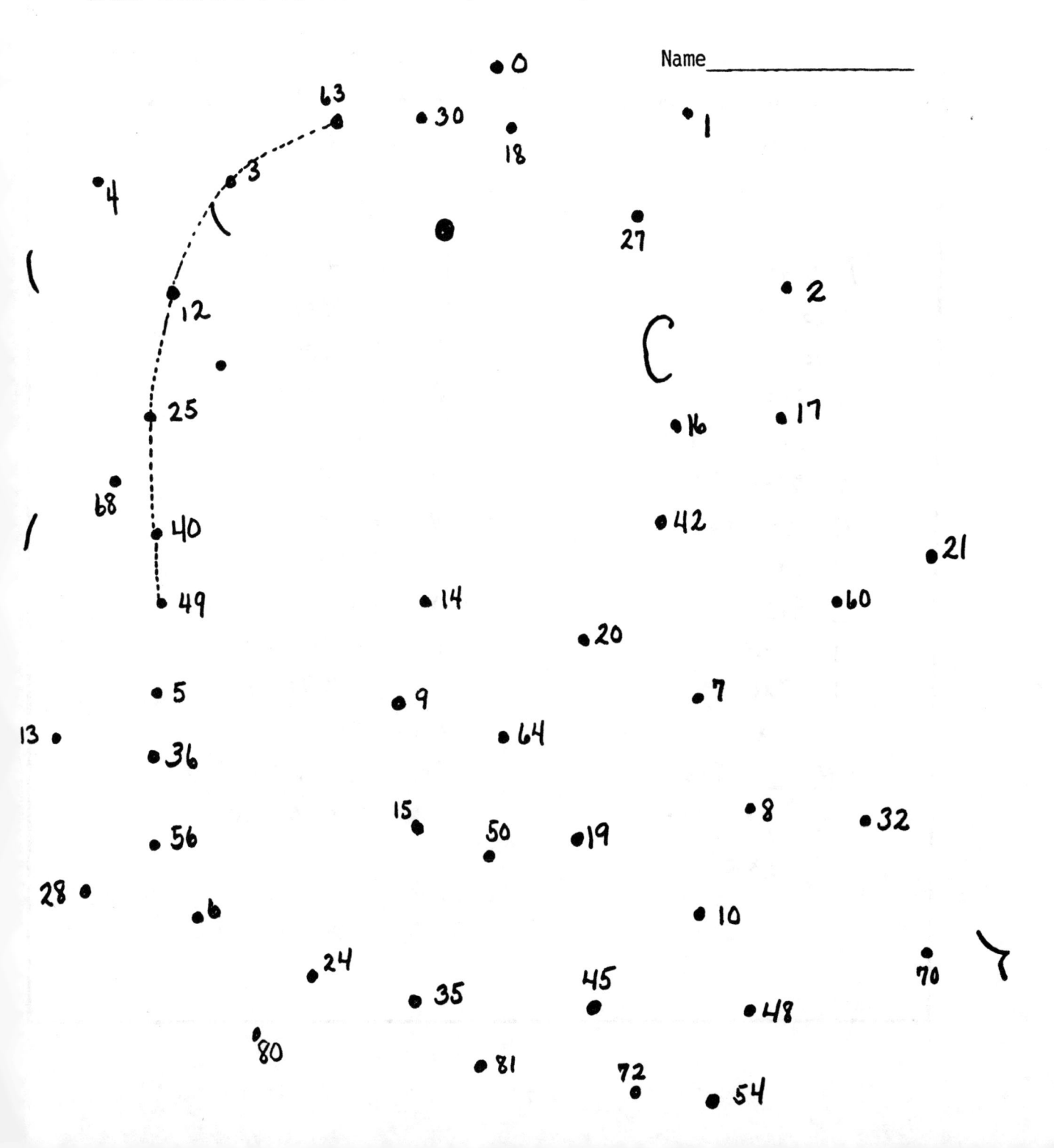

Name_____

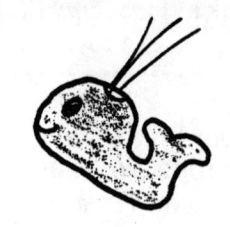

# Directions:

Solve A. Find the answer in your dot picture. Start there.
Solve B. Draw a line from the A answer to the B answer.
Continue until your picture is complete.
Try a picture of your own.

A  7×7=
B  5×8=
C  5×5=
D  2×6=
E  __ × 8 = 24
F  7×9=
G  2 × __ = 60
H  3×6=
I  9×3=
J  4×4=
K  7×6=
L  5×4=
M  __ × 7 = 63
N  3 × __ = 45
O  5×10=

P  1×19=
Q  __ × 8 = 56
R  4 × __ = 32
S  __ × 10 = 100
T  6×8=
U  9×6=
V  8×9=
W  5×9=
X  5×7=
Y  6×4=
Z  __ × 9 = 54
AA  7×8=
BB  __ = 6×6
CC  __ × 20 = 100

PAIL O' PROBLEMS

---
Skill: *Writing and solving related multiplication and division equations*
---

1. Put the following materials into a plastic pail:

    --a deck of cards containing two of each factor 0-10 (written in one color ink)

    --a deck of cards containing two of each product 0-100 (written in a second color ink)

    --a deck of cards containing two of each symbol: x, =, and ÷

    --a four-foot length of clothesline

    --about 20 clothespins

2. The directions may be written on the pail:

    Play this game with a partner!
    1. One player 'hangs' a multiplication problem.
    2. The other player must 'hang' a division problem using the same numbers.
    Do several problems this way. Sometimes try the division problem first!

Variation: The same game works well for addition and subtraction!

# GRAB BAG

---
**Skill:** *Writing multiplication and division equations to solve story problems*

---

1. Fill two grab bags (or large envelopes) with bits of information that might be used in problems.

    Examples:

    15¢ each         89 boxes
    2 friends        27 houses
    6 bottles        3 cartwheels
    45 chairs        9 gorillas

2. In a third bag, place several cards saying either MULTIPLY or DIVIDE.

3. A student must draw one card from each bag and write a story problem using the information he has chosen. His third card will tell which operation to use in his problem. If the two cards drawn cannot be used together, he may draw again.

   Variation: For more challenging problems, add a third bag of facts so that students are required to use all three.

# PATCHWORK SNAKE

---
**Skill:** *Estimating missing factors, products, and quotients*
---

1. Have students sew together scraps of fabric about a foot long each to make a long walk-on "snake."

2. On each section pin or write an equation (with a factor, product, or quotient missing).

3. To play the game, all players begin on START. A non-playing official throws one die to determine the number of patches each player moves.

4. The player must estimate the answer to each problem on which he lands. (The "official" holds an answer sheet to check the estimate given.)

5. If a player's estimate is not close to the answer, he must remain on that spot until his next move when he tries again.

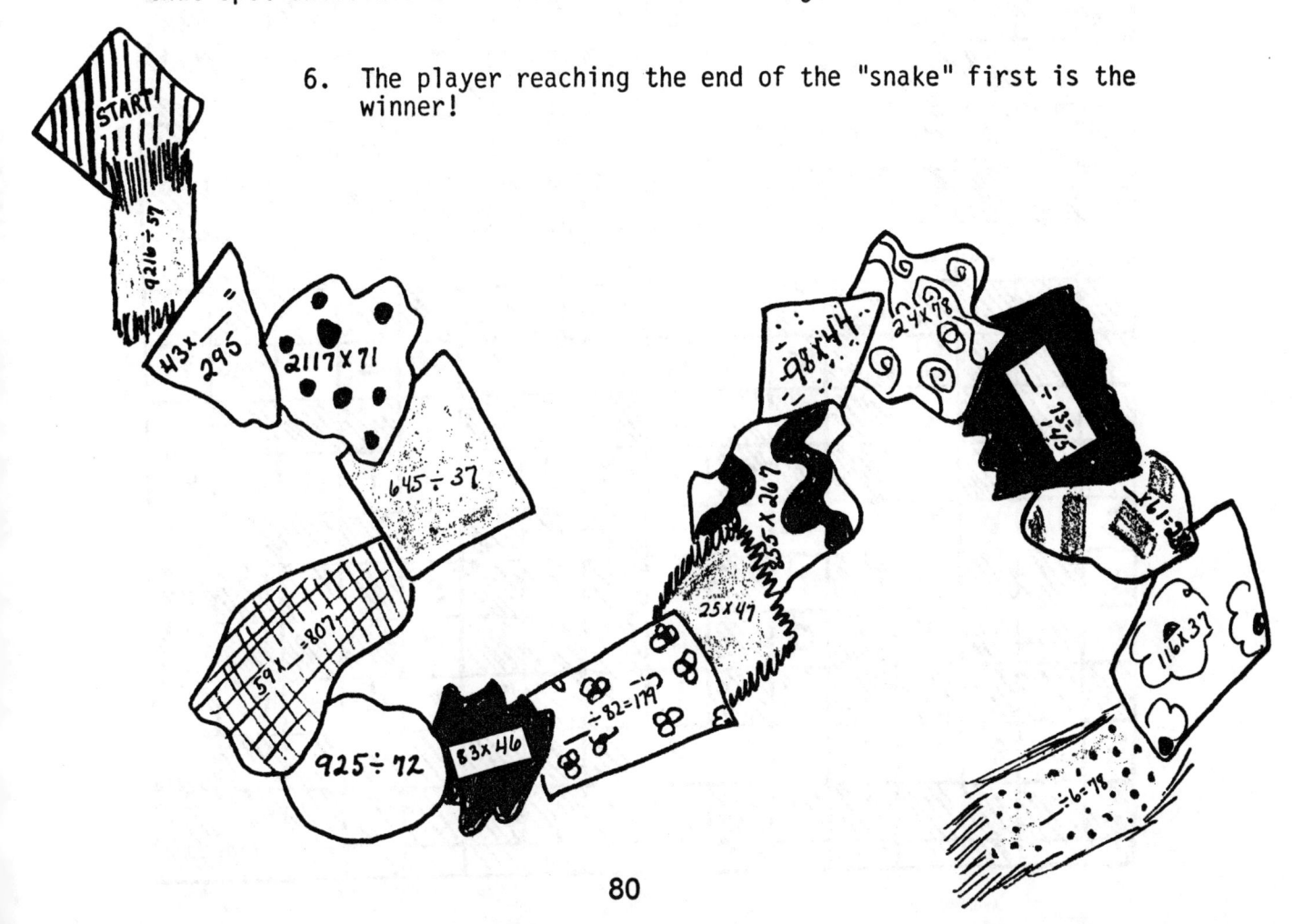

6. The player reaching the end of the "snake" first is the winner!

TERM SQUIRM

---
Skill: *Defining and using terms associated with the four basic operations*
---

1. Make a large crossword puzzle on a long strip of mural paper.

2. Cover it with clear adhesive paper so that students may write on it and hang it on a wall.

   Variation: Duplicate the puzzle on the following page so that each student has his own copy of the puzzle to work.

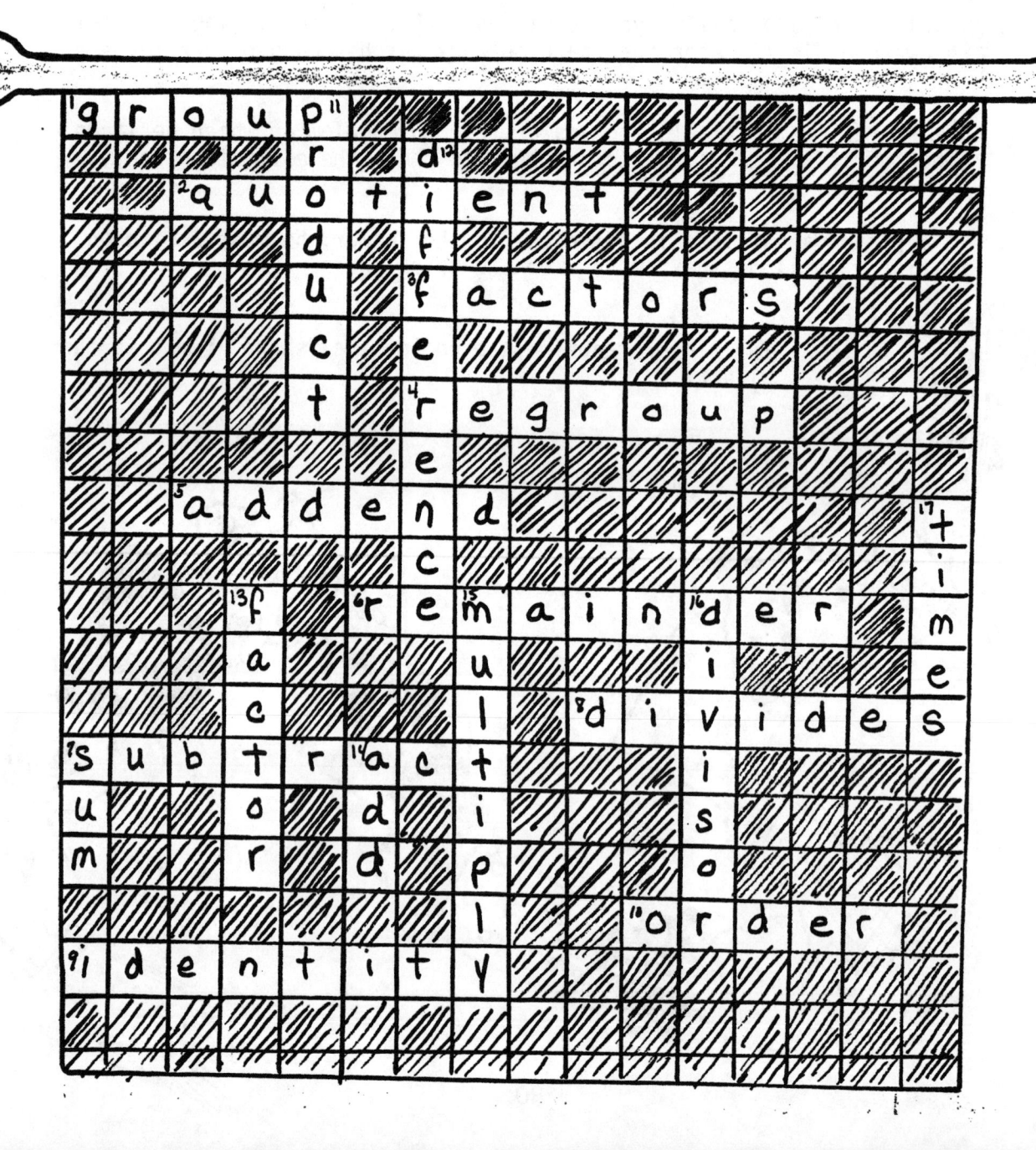

# TERM SQUIRM

Name _____

## Across

1. The Associative Property has to do with how you _____.
2. The answer in a division problem.
3. You multiply these.
4. Sometimes, when adding or subtracting, you must _____.
5. A number that you add is called an _____.
6. This is the 'left over' in division.
7. When you find the difference, you _____.
8. When Mary does this problem (20÷6) she _____.
9. Zero is the _____ element for addition.
10. It doesn't matter what _____ you add addends.

## Down

?. The answer in addition.
11. The answer in multiplication.
12. The answer in subtraction.
13. In 8×4, 8 is a _____.
14. To solve this (7+9) you _____.
15. To find a product, you _____.
16. The number you divide by.
17. Four _____ nine equals thirty-six.

82

# LADYBUG LEAP

**Skill:** *Using a number line to solve simple multiplication and division equations*

1. Use a long roll of shelf paper (preferably vinyl) to make a floor number line. Number from 0-100.

2. Make ten ladybugs: one with one spot, one with two spots, one with three spots, and so on through ten. Staple or glue each bug to a popsicle stick.

3. The number line can be used to:

    a. show multiplication as repeated addition and division as repeated subtraction.

    b. introduce the concept of factors and divisors.

    c. begin solving division problems with remainders.

4. All ladybugs begin at zero. At each turn, a ladybug can only fly the same number of spaces as she has spots.

5. Play a game by asking some of these kinds of questions:

    --*How many jumps will the 6 bug have to take to reach 24?*

    --*Can the 4 bug land on 15?*

    --*If the 7 ladybug hops seven times, where will she land?*

    --*Can the 9 bug begin at 72 and, hopping backward, land on zero?*

    --*How many backward hops from 56 can the 8 bug make?*

    --*If Ladybug Six moves backward from 45, how many hops can she make? What will be left over?*

6. Put several multiplication and division problems into a large "ladybug" envelope. Students may find the answers using the number line.

## JUST AVERAGE!

---
Skill: *Finding the average of a set of numbers*
---

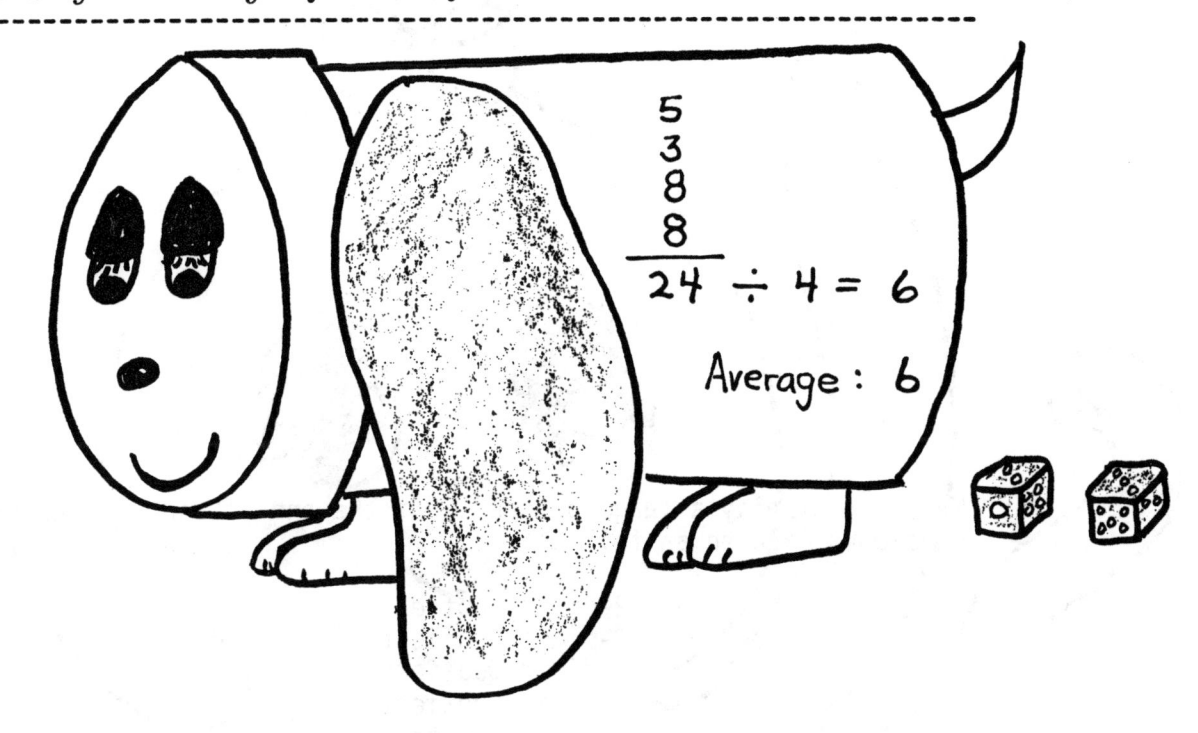

$$\begin{array}{r} 5 \\ 3 \\ 8 \\ 8 \\ \hline 24 \div 4 = 6 \end{array}$$

Average: 6

1. Use an oatmeal box to create a portable "Just Average" kit. Make the box into a dog or any other object you wish. Include an example of how to find averages.

2. Inside the box put:  a pair of dice
   a 12" x 12" piece of felt
   exercise cards

3. Students toss the dice on the mat a given number of times, total their results, and divide to find the average.

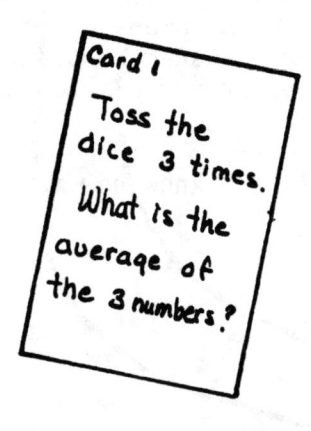

> **Card 1**
> Toss the dice 3 times. What is the average of the 3 numbers?

## DEEP SEA DIVISION

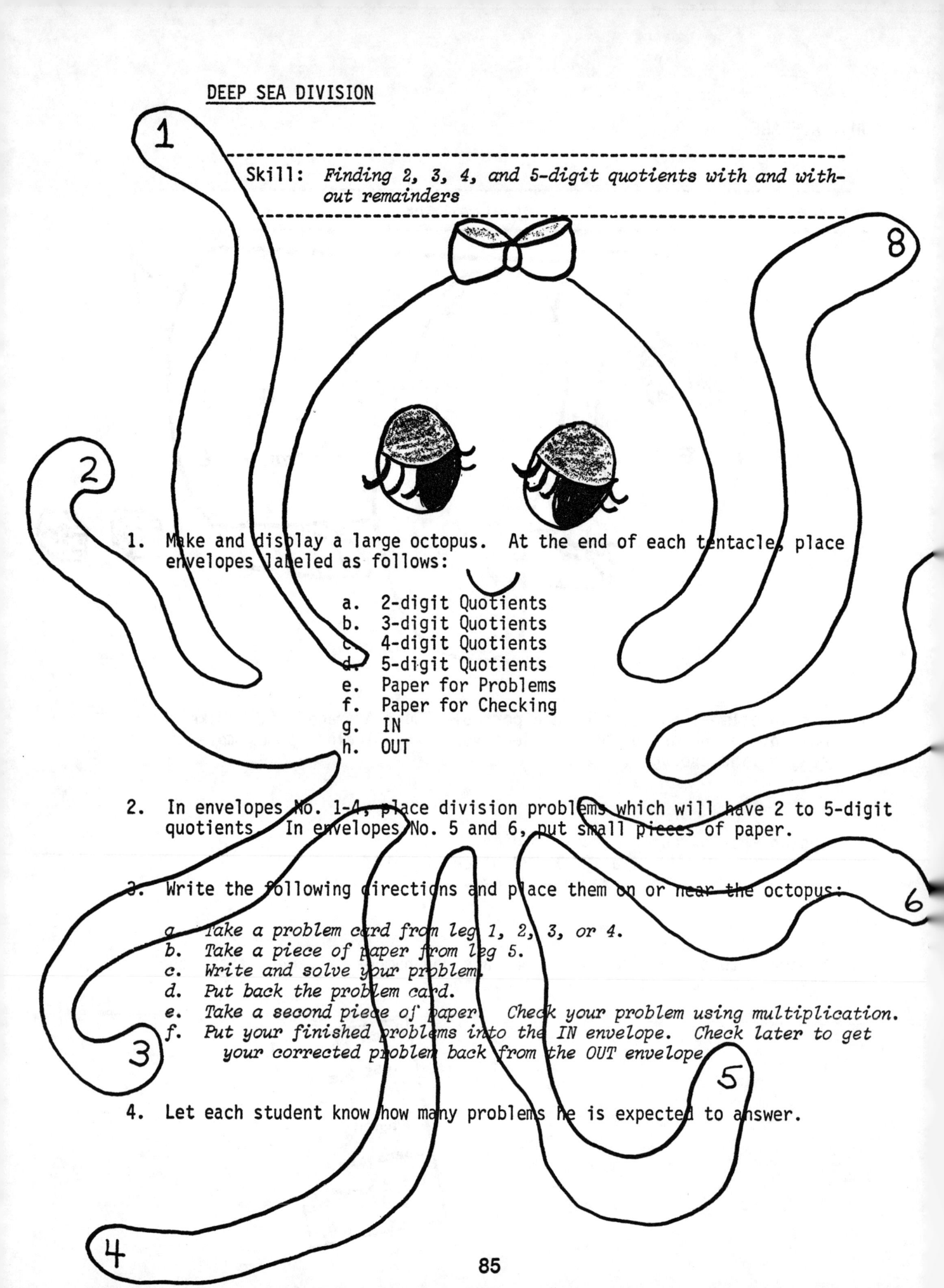

**Skill:** *Finding 2, 3, 4, and 5-digit quotients with and without remainders*

1. Make and display a large octopus. At the end of each tentacle, place envelopes labeled as follows:

    a. 2-digit Quotients
    b. 3-digit Quotients
    c. 4-digit Quotients
    d. 5-digit Quotients
    e. Paper for Problems
    f. Paper for Checking
    g. IN
    h. OUT

2. In envelopes No. 1-4, place division problems which will have 2 to 5-digit quotients. In envelopes No. 5 and 6, put small pieces of paper.

3. Write the following directions and place them on or near the octopus:

    a. Take a problem card from leg 1, 2, 3, or 4.
    b. Take a piece of paper from leg 5.
    c. Write and solve your problem.
    d. Put back the problem card.
    e. Take a second piece of paper. Check your problem using multiplication.
    f. Put your finished problems into the IN envelope. Check later to get your corrected problem back from the OUT envelope.

4. Let each student know how many problems he is expected to answer.

## ROLL A SENTENCE

**Skill:** *Writing number sentences using more than one operation*

1. In a box, place four dice and a stack of light-colored construction paper. Include a small piece of felt for rolling dice quietly.

2. Inside the box, write the following directions:

> Take one die and a piece of paper.
> Roll 4 different numbers.
> Use all 4 numbers to write a number sentence.
> Write 5 more sentences by doing the same thing 5 more times.
> * Try rolling 6 numbers.
> * Try rolling 2 dice three times each.
> * Ask a friend to race you. Roll the dice and see who can write a sentence first.

## OUTCASTS

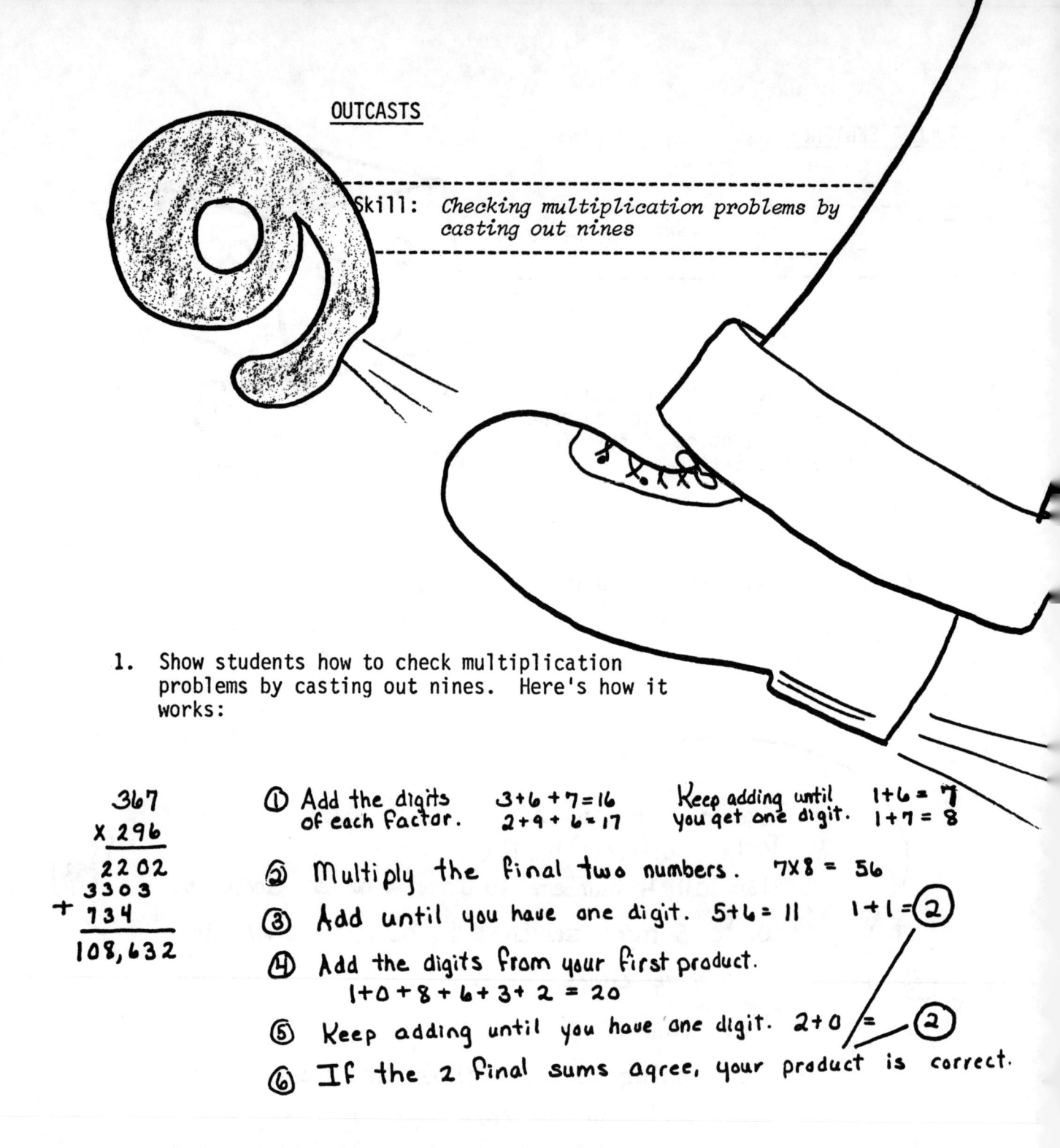

**Skill:** *Checking multiplication problems by casting out nines*

1. Show students how to check multiplication problems by casting out nines. Here's how it works:

```
   367
 x 296
  2202
  3303
+  734
───────
108,632
```

① Add the digits of each factor.   3+6+7=16    Keep adding until   1+6 = 7
                                    2+9+6=17   you get one digit.  1+7 = 8

② Multiply the final two numbers.  7x8 = 56

③ Add until you have one digit.  5+6 = 11    1+1 = ②

④ Add the digits from your first product.
   1+0+8+6+3+2 = 20

⑤ Keep adding until you have one digit.  2+0 = ②

⑥ If the 2 final sums agree, your product is correct.

2. Prepare and duplicate a sheet of multiplication problems. Leave room for checking, using this method.

3. Ask students if they can discover why this method is called "casting out nines" and how it works?

# SOME LIKE IT HOT . . .

---
Skill: *Reading positive and negative integers on a thermometer*

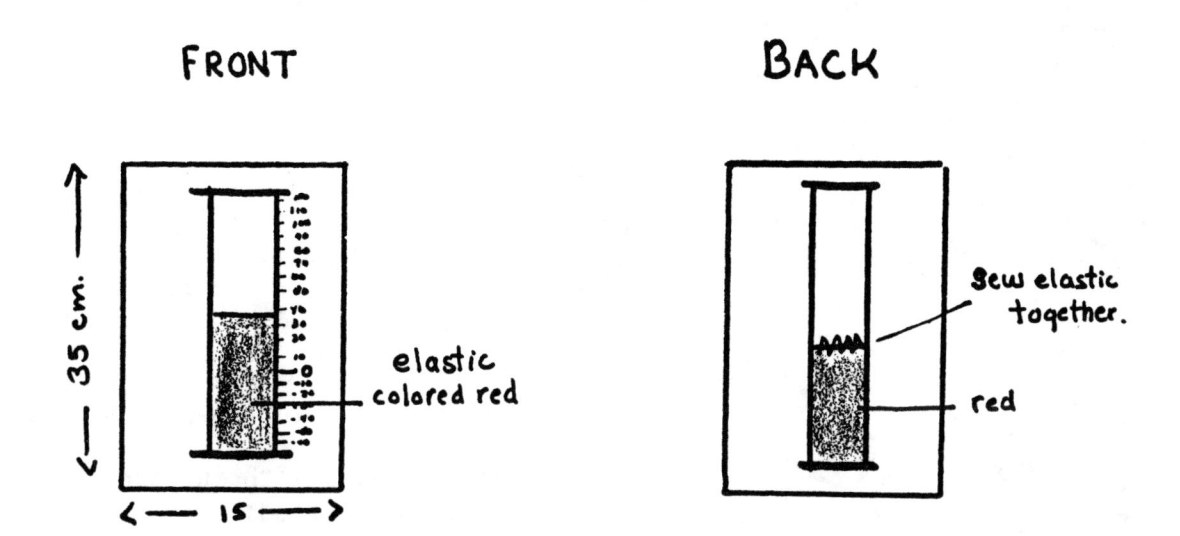

---

**FRONT**  **BACK**

1. Have students make personal thermometers using cardboard, magic marker, and one-inch-wide elastic.

   ### HOW TO MAKE A THERMOMETER

   Cut a two-foot piece of elastic. Color half of the piece red. Slit the cardboard 4 cm. from the top and bottom. Insert the elastic pieces and sew or staple the ends together in the back. Draw markings along the right side of the thermometer at intervals of 1 cm. Begin numbering with zero. Continue up and backward at intervals of 10°. Mark ten spaces (1 mm. each) between each interval.

2. Give oral and written exercises which require students to show various temperatures on their thermometers:

   *Show 27 degrees below zero.*
   *Show 55 degrees.*
   *Close your eyes and move the elastic. Open . . . what is the temperature?*
   *Set a temperature. Show it to a friend. Have him read the temperature. Is he correct?*

   Variation: You'll want to save these thermometers for exercises with adding and subtracting integers.

## OFF TO THE RACES

**Skill:** *Adding and subtracting positive and negative integers*

1. Make a long number line on tagboard (six feet long). Number it with positive and negative integers. Write START at zero and FINISH at the far right end.

2. Also needed for the Turtle-Hare Race are:

   a. a deck of cards with integers to be added or subtracted:

   i.e:  + (+4), + (-3)

   - (-7), - (5)

   b. a turtle
   c. a hare
   d. two players

3. Give the students the following directions for this race:

   --*One player is the turtle; one is the hare.*
   --*Both start at zero.*
   --*When it is your turn, draw a card and do what it says.*
   --*The player whose animal reaches FINISH first is the winner.*

FOOTBALL FRENZY

---
Skill: *Adding and subtracting integers*

---

1. Transform a long, shallow box into a football field. Have students cover the bottom inside with green paper or felt; make goalposts from pipe cleaners or paste sticks; draw and number the yard lines and attach a plastic-covered scoreboard to each end.

   Markers or miniature players may be made to represent the teams.

2. Make about 50 PLAY BALL cards like the ones below:

   | Interception: Lose the ball | Short Run: Advance 4 yards | Quarterback hit behind the line: Lose 10 yards | Long Pass: Advance 35 yards |

3. Make 20 EXTRA POINTS cards similar to these:

   | What is $-13 + -16$? | $\_ + -6 = -4$ | What is $7 - (-5)$? |

4. Inside the boxtop, write these RULES OF THE GAME:

   a. *Make or choose a marker to represent your team.*

   b. *Roll the dice. The team with the lowest number must kick off to the other.*

   c. *The receiving team draws a PLAY BALL card for each of four downs. If a first down is reached, they may draw four more.*

   d. *If no first down is made, the ball goes to the opposing team.*

   e. *If a team is within 25 yards of the goal on the fourth down, they may try for a field goal by drawing an EXTRA POINTS card and answering the question.*

   f. *After a touchdown, make an extra point by drawing and answering correctly.*

   g. *Each team must keep a written list of its moves.*

   h. *Keep score: 6 points per touchdown; 3 points for field goal*

   i. *Set the timer: 10 minutes for each half.*

5. Label the outside, put all the equipment inside (timer, score cards, "players," dice, sample tally sheet) and the game is ready to go!

**bright idea!**

How big is a million? Have students collect 1,000,000 bottle caps, corn kernels, or grains of rice, or ask them to figure:

   how many trips around the world to make 1,000,000 miles.
   how many days to make 1,000,000 minutes.
   how many math books to make 1,000,000 pages.
   how many 70 pound kids to make 1,000,000 pounds.
   how many dozens to make 1,000,000 eggs.

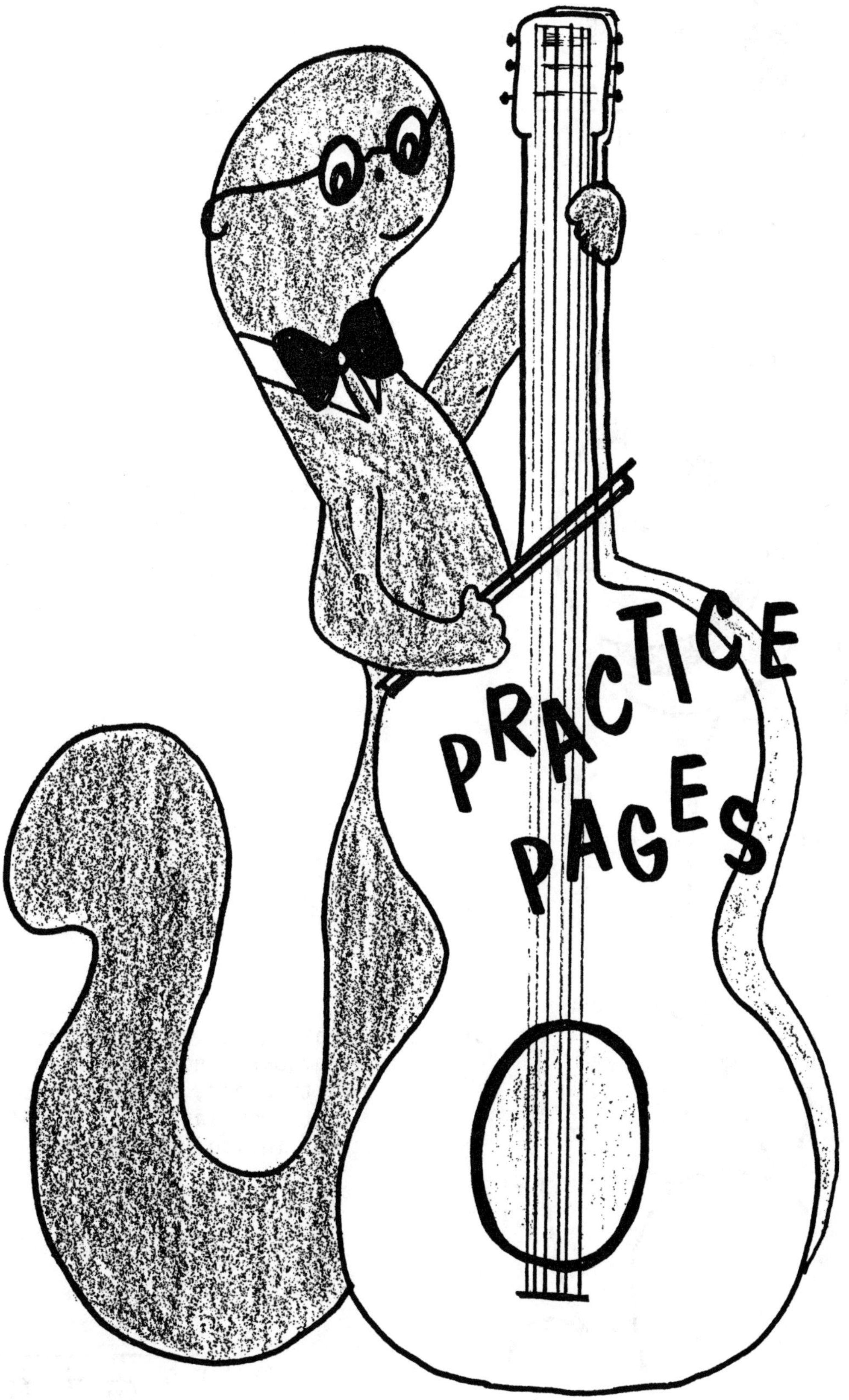
PRACTICE
PAGES

# NO MONKEY BUSINESS

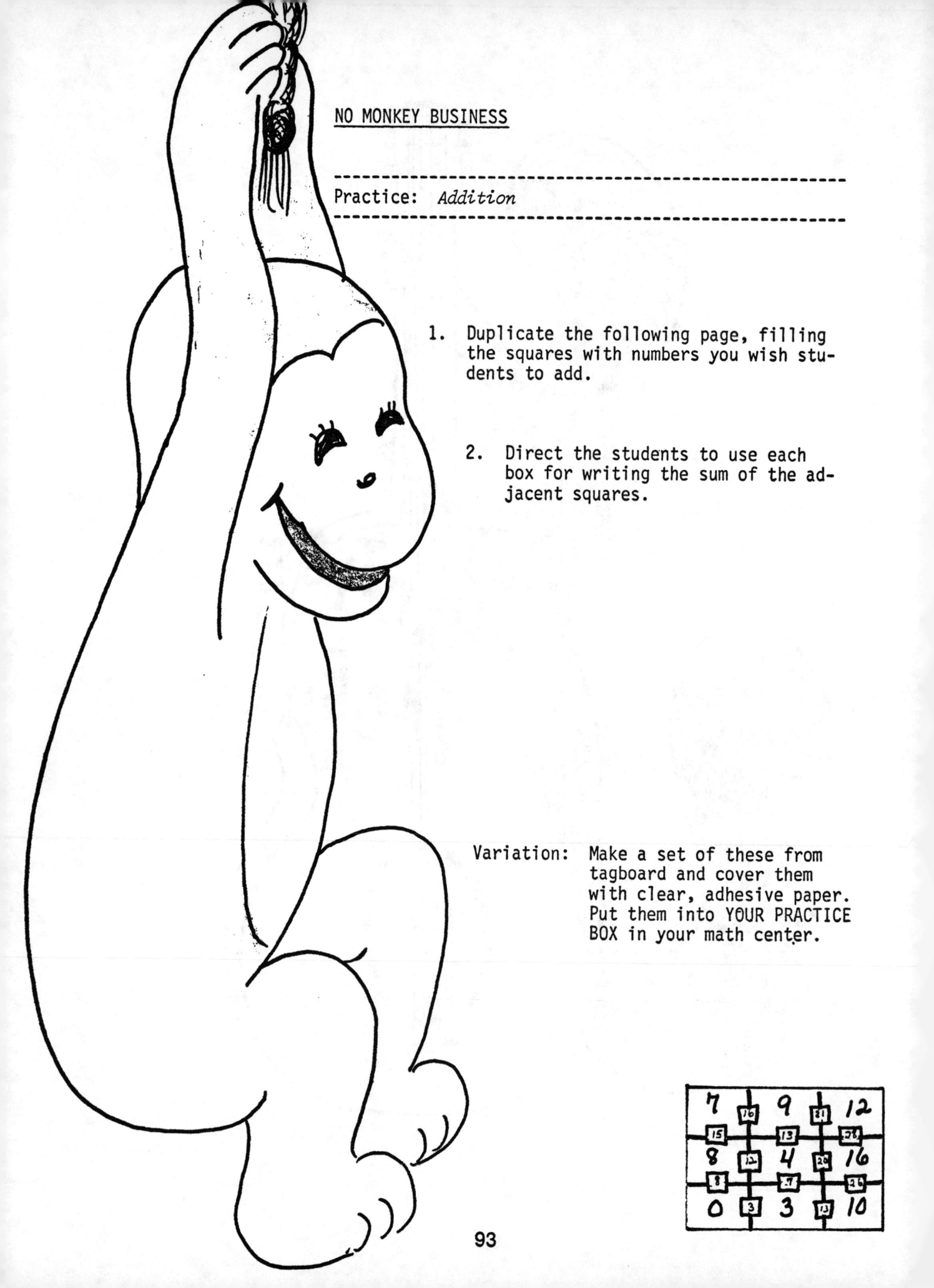

Practice: *Addition*

1. Duplicate the following page, filling the squares with numbers you wish students to add.

2. Direct the students to use each box for writing the sum of the adjacent squares.

Variation: Make a set of these from tagboard and cover them with clear, adhesive paper. Put them into YOUR PRACTICE BOX in your math center.

93

Name _____

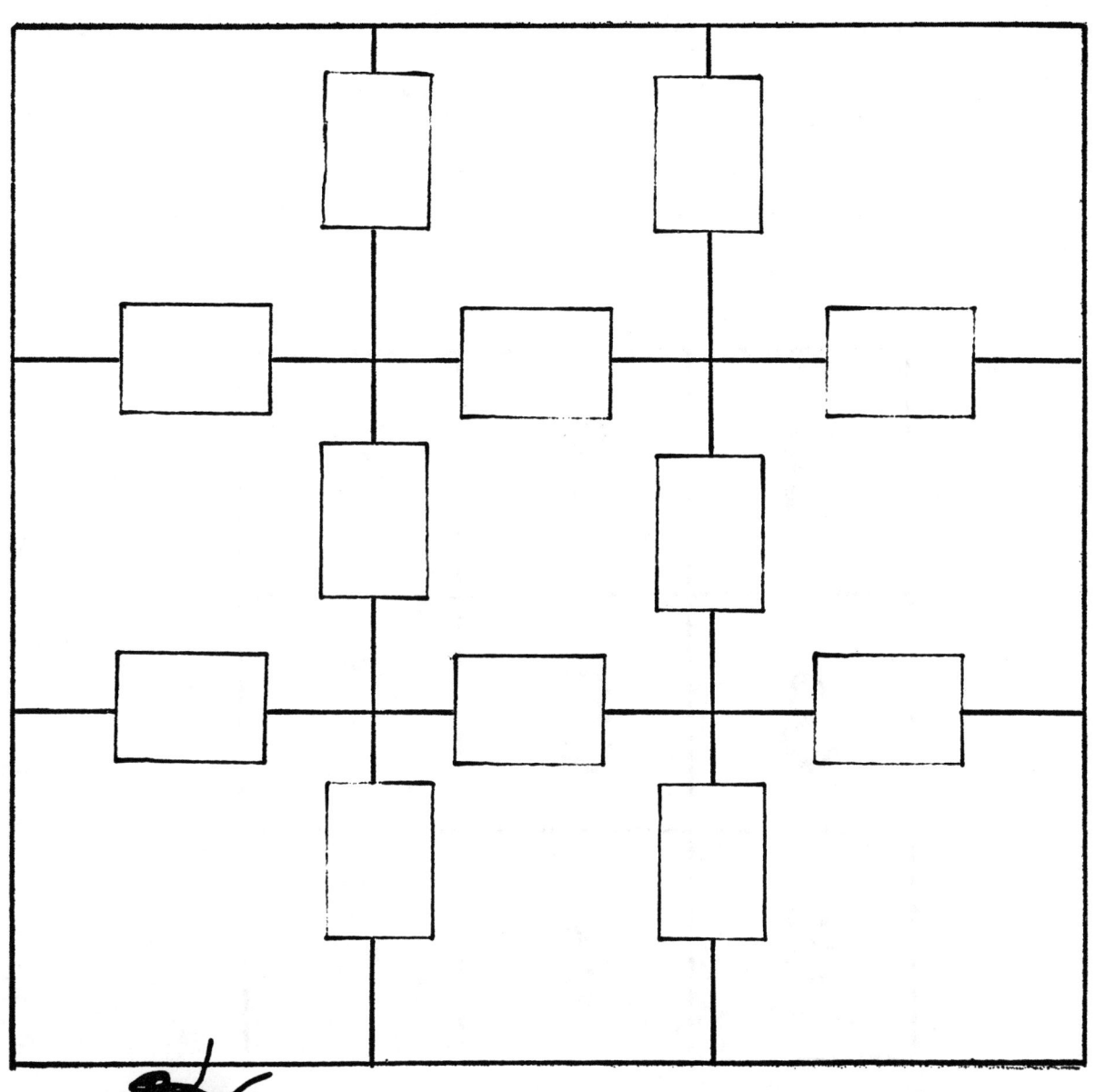

In each ☐ write the sum of the numbers in the two adjoining boxes.

94

# MAGIC SQUARES

---
Practice: *Addition*
---

1. Show students a magic square. See if they can discover why it is magic. (The sum of all the rows is the same.)

2. Duplicate a page of squares. Ask the students to determine which ones are magic.

|   |   |   |
|---|---|---|
| 4 | 9 | 2 |
| 3 | 5 | 7 |
| 8 | 1 | 6 |

3. Provide another worksheet containing "unfinished" squares. Students fill in the missing numbers to make a magic square.

Hint: You can make more magic squares from this one! Just add the same number to each digit, and you have a brand new one!

| 6  | 11 | 4 |
|----|----|---|
| 5  | 7  | 9 |
| 10 | 3  | 8 |

95

NUMERALS NELDA 'N NORMAN
-------------------------------------------------------------------
Practice: *Addition*
-------------------------------------------------------------------

1. Make a "Numeral Nelda" or "Norman" or other numeral person.

2. Let students discover his/her age by finding the sum of all the numbers used to make him/her.

$$\begin{array}{r} 3 \\ 3 \\ 3 \\ 8 \\ 9 \\ 9 \\ 0 \\ 6 \\ 7 \\ + 11 \\ \hline \end{array}$$

3. Ask the students to make their own numeral persons. Then they may trade with friends to find the ages.

4. Encourage students to try other creatures too!

5. Find a place for the creatures to be displayed. Pictures clipped to a hanger or line can be easily used and returned by students needing addition practice.

6. The artist for each can keep the answer so that anyone solving his Problem has to visit him to check the answer.

# A-MAZE-ING!

------------------------------------------------------------
Practice: *Column addition*
------------------------------------------------------------

1. Make a large maze similar to the one shown. Draw it on an old sheet or other strip of fabric (about 4' x 4'). Make it more attractive by coloring some spaces or by gluing on cut pieces of brightly-colored felt.

2. Explain to the students that only one column has the proper sum! In order to find which it is, they must add each column.

SEARCH 'N CIRCLE

---
Practice: *Addition and subtraction facts*
---

Give copies of this number puzzle to students. Ask them to follow the written directions.

## Search 'n Circle!

| 4 | 7 | 11 | 9 | 2 | 8 | 10 | 4 | 14 |
|---|---|---|---|---|---|---|---|---|
| 9 | 6 | 15 | 8 | 7 | 6 | 13 | 4 | 9 |
| 13 | 1 | 12 | 3 | 9 | 2 | 7 | 8 | 5 |
| 2 | 10 | 12 | 5 | 7 | 9 | 11 | 6 | 4 |
| 11 | 8 | 19 | 3 | 16 | 1 | 18 | 2 | 1 |
| 5 | 1 | 6 | 10 | 12 | 8 | 4 | 17 | 12 |
| 6 | 8 | 14 | 4 | 10 | 6 | 4 | 9 | 13 |
| 6 | 3 | 9 | 2 | 12 | 9 | 3 | 8 | 8 |
| 12 | 4 | 16 | 8 | 8 | 15 | 10 | 5 | 5 |

Circle 10 addition equations.
Circle 10 subtraction equations.
Can you find more?

## HALLOWEEN + EASTER = ST. VALENTINE'S DAY

---
Practice: *Addition and subtraction*
---

1. Show your students how to add and subtract words and sentences.

```
  Halloween      9 letters
+   Easter     + 6 letters
     15
```

2. Give them lots of problems to solve, i.e:

```
  Cotton candy
-   popcorn
```

```
  Tomorrow is Saturday.
+ Did you bring crayons?
```

3. They may make their own problems using their names or information about themselves.

```
    Jennifer
+   Redding
```

4. As an extra challenge, ask students to find a word or sentence as the sum, instead of a number.

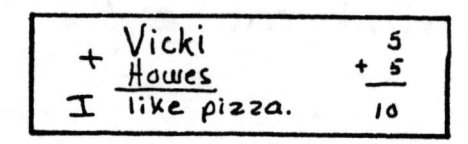

```
+    Vicki         5
     Howes      + 5
  I like pizza.   10
```

99

# TUG-OF-WAR

Practice: *All basic operations with whole numbers*

1. Two groups or classes may have a tug-of-war to see which side has more "strength with numbers."

2. Give the numerals 0-100 to each group.

3. Explain these rules:

    a. *Use the numerals to make as many equations as you can.*
    b. *Use any of the four operations.*
    c. *Use all the numerals.*
    d. *Each addition equation is worth 2 points.*
    e. *Each subtraction equation is worth 4 points.*
    f. *Each multiplication equation is worth 6 points.*
    g. *Each division equation is worth 8 points.*
    h. *Mixed equations are worth 10 points each.*

4. Give each group a limited amount of time (about 15 minutes). At the end of the time period, check equations for accuracy, tally the scores, and admire the "muscles" of the winner!

Variation: Use a greater variety of math skills letting students form inequalities, decimals, fractions, integers, etc.

DIAL-A-DOUGHNUT

------------------------------------------------------------
Practice: *Long multiplication and division*
------------------------------------------------------------

1. Cut cardboard doughnuts of two sizes (10 in. in diameter and 15 in. in diameter). Cut a small hole in the center of each doughnut.

2. On the different circles, write numbers of various digit length:

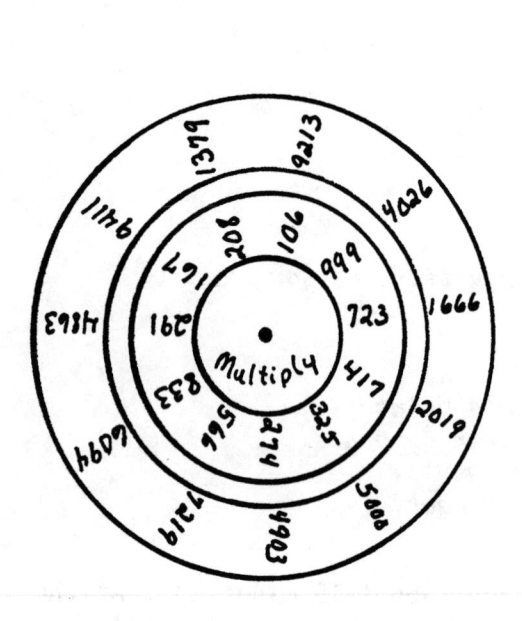

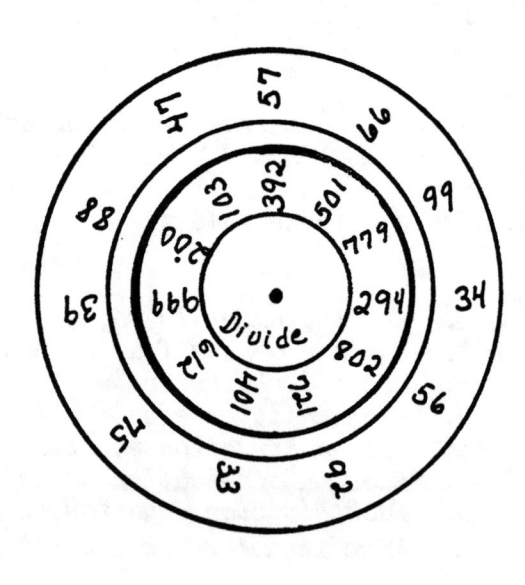

3. Students place a small and a large doughnut together over a spindle or peg, and write multiplication or division problems from the combination of numbers. Loads of problems are possible using each pair of doughnuts!

NAPIER'S BONES

---
Practice: *Basic multiplication facts*
---

*Napier's Bones are portable multiplication rods. The index bone lists the factors 1-9. Each of the other bones contains the multiples of one of those nine numbers.*

1. Have each student make a set of ten Napier's bones from tongue depressors or bone-shaped pieces of cardboard.

2. As a group, practice using the bones. To multiply 3 x 5, use the index bone and the 5 bone like this:

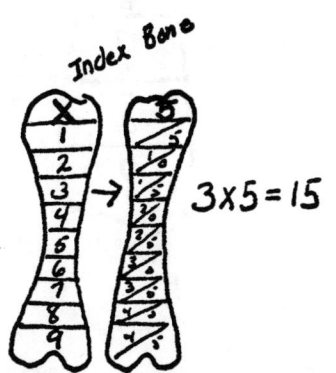

   3 x 5 = 15

102

To divide 72 by 8, use the index bone and the 8 bone like this:

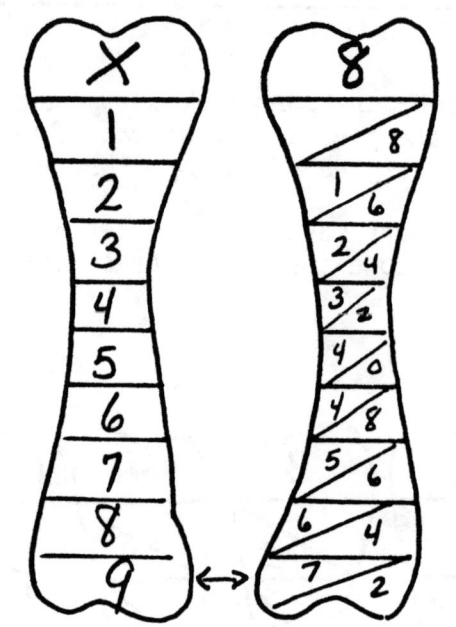

Find 72 on the 8-bone. Place it next to the Index Bone. 72 lies next to the 9, so 72 ÷ 8 = 9.

To multiply more than one digit, use the bones this way:

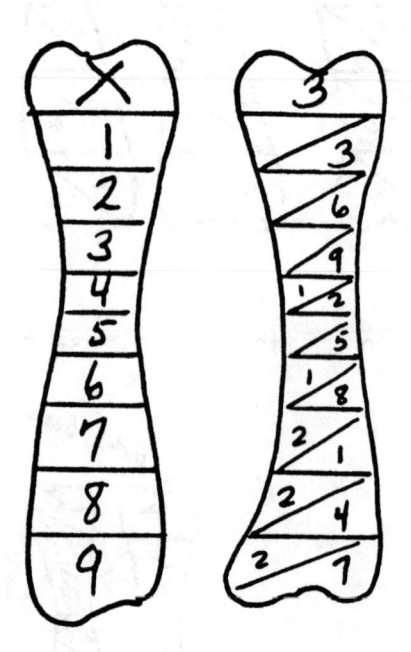

$$\begin{array}{r} 847 \\ \times \phantom{00}3 \\ \hline \end{array}$$

3 × 7 =  21
3 × 4 =  12
3 × 8 =  24
―――――
        2541

## KEEP IN SHAPE!

---
Practice: *Basic multiplication facts using factors 1-9*
---

1. Give students a copy of the following:

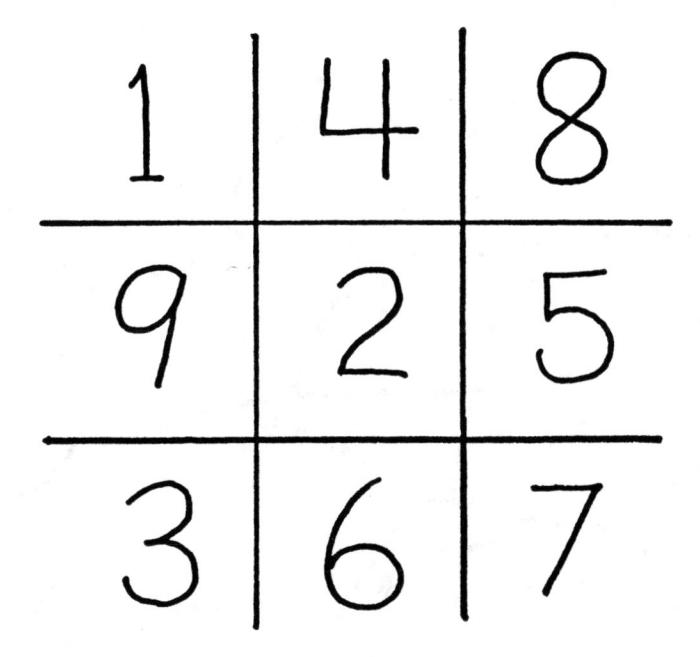

   | 1 | 4 | 8 |
   |---|---|---|
   | 9 | 2 | 5 |
   | 3 | 6 | 7 |

2. Ask them to notice the shape of the "container" in which each factor has been placed.

3. Duplicate several problems for them to solve using the factors on the grid:

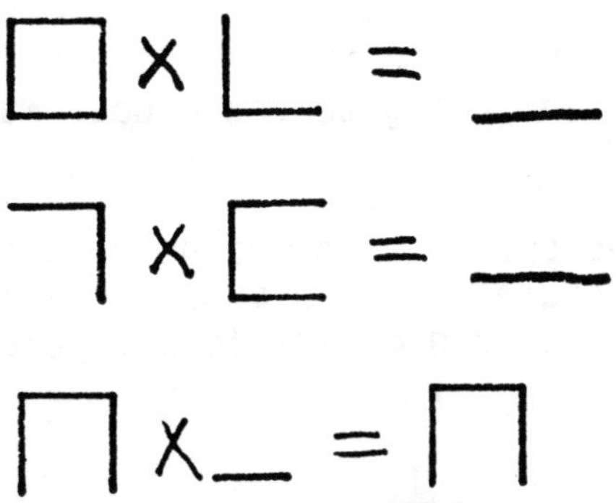

   □ × L = __

   ⊓ × ⊏ = __

   ⊓ × __ = ⊓

## IN LIVING COLOR

Practice: *Basic multiplication and division facts*

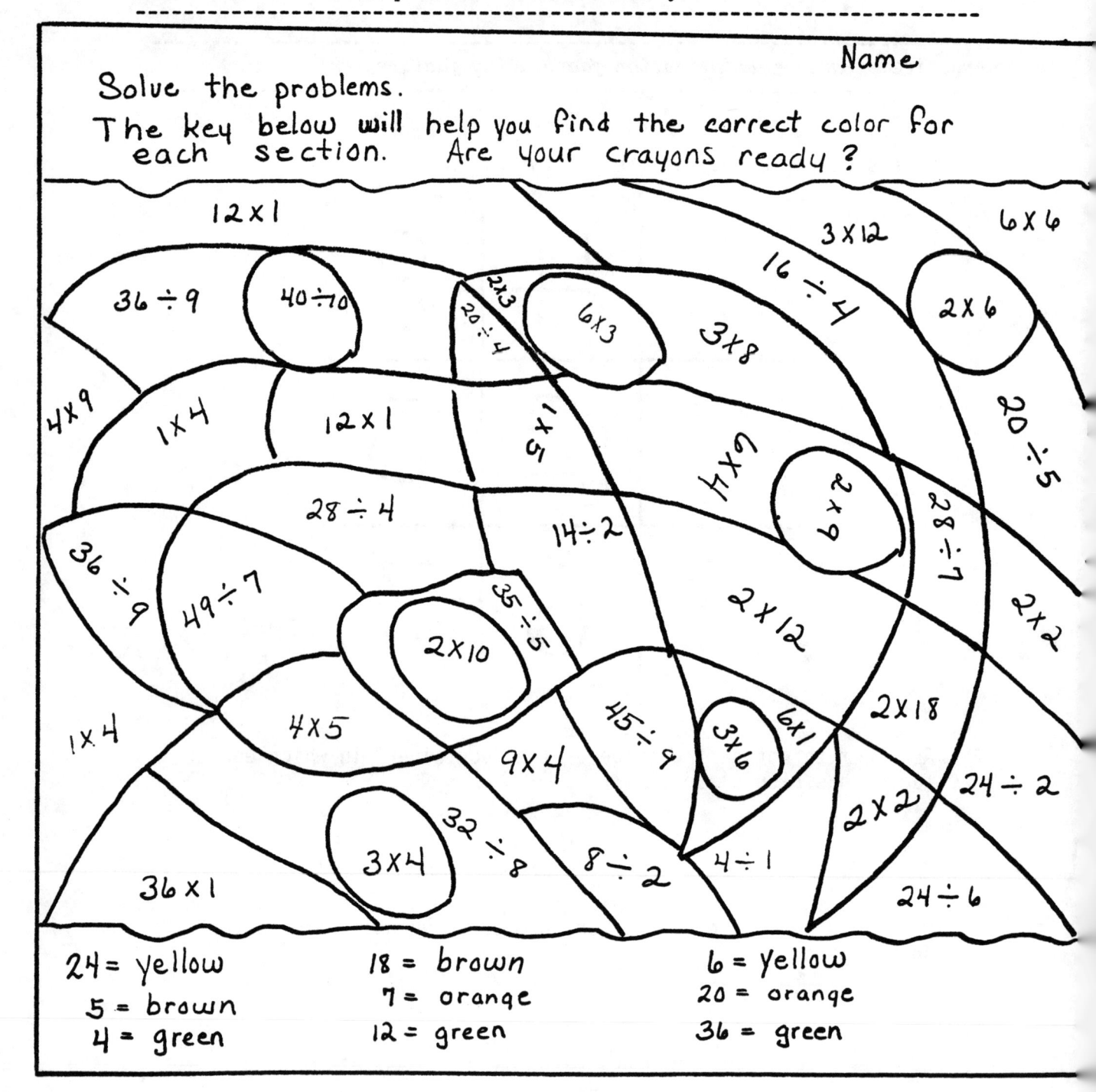

Name _____

Solve the problems.
The key below will help you find the correct color for each section. Are your crayons ready?

*(puzzle sections with problems:)*

12×1, 3×12, 6×6, 36÷9, 40÷10, 2×3, 6×3, 3×8, 16÷4, 2×6, 4×9, 1×4, 12×1, 20÷4, 5×1, 4×6, 20÷5, 28÷4, 14÷2, 2×9, 28÷7, 36÷9, 49÷7, 35÷5, 2×12, 2×2, 2×10, 45÷9, 3×6, 6×1, 2×18, 1×4, 4×5, 9×4, 2×2, 24÷2, 3×4, 32÷8, 8÷2, 4÷1, 24÷6, 36×1

**Color key:**

- 24 = yellow
- 5 = brown
- 4 = green

- 18 = brown
- 7 = orange
- 12 = green

- 6 = yellow
- 20 = orange
- 36 = green

This puzzle becomes a picture when the problems are solved correctly and each section is colored according to the code. Make one of your own or duplicate this one.

Variation: A mural-sized puzzle with dozens of problems forming a complex picture can be a group project that challenges everyone (and offers lots of practice)!

# LATTICE WORK

---
Practice: *Long multiplication*
---

1. Lattice multiplication is fascinating--and it's a fun, painless way to check long multiplication problems. Teach your students how to do it!

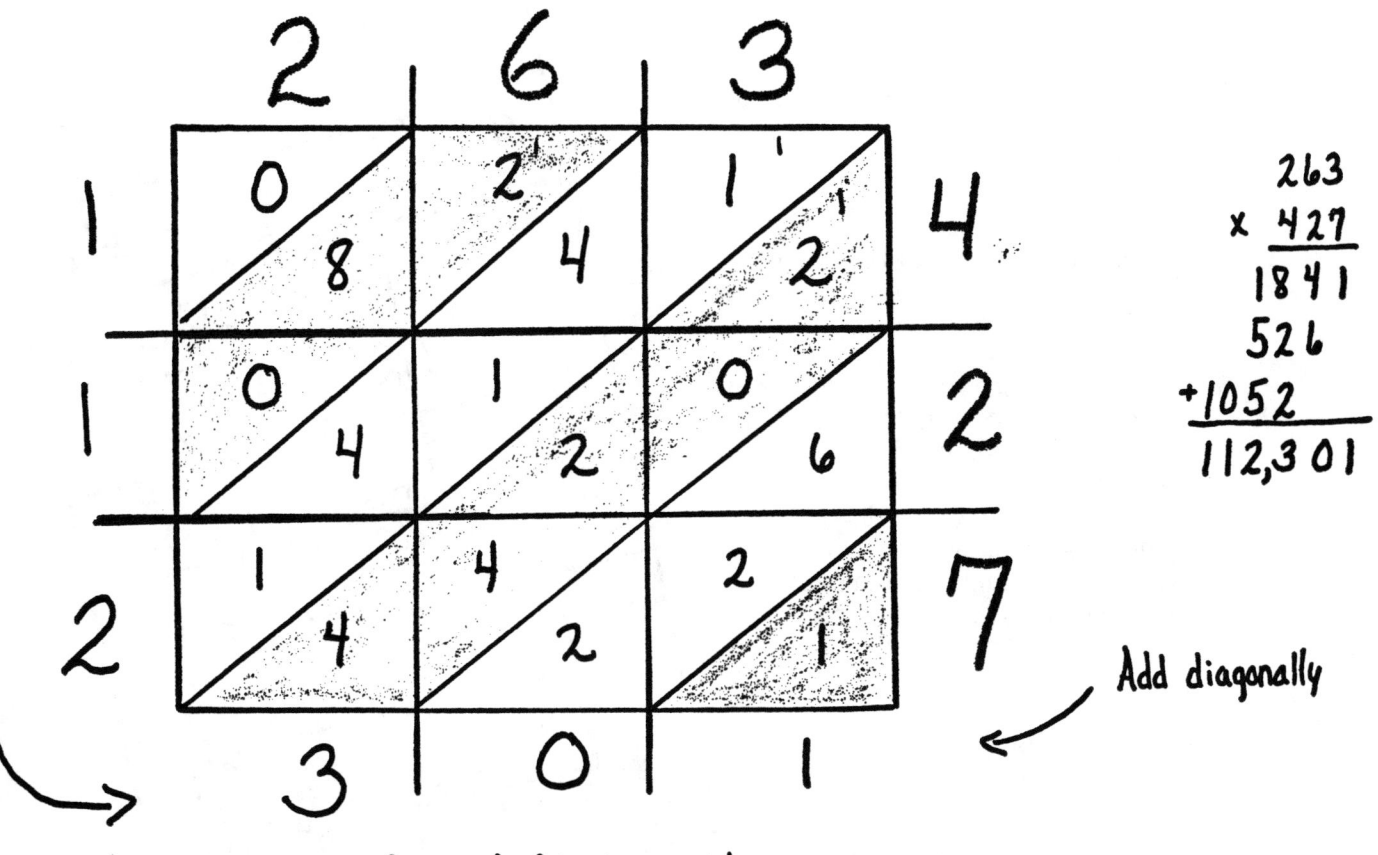

```
      263
    x 427
     1841
      526
    +1052
    112,301
```

Add diagonally

Read answer from left to right.  112,301

2. Make a few lattice grids covered with clear plastic so that they can be used again and again.

106

TOUCH PUZZLES

---
**Practice:** *Basic addition, subtraction, multiplication and division facts*
---

1. Paste a large picture (or draw one) on a piece of cardboard.

2. On the back of the board, draw cutting lines to make the picture into a puzzle.

3. In each section write a problem. Write the answer to that problem along the edge of the touching piece.

4. Cut apart the puzzle pieces and place them in a large envelope.

5. A student solves the problems and puts the puzzle together with the picture facing down. Working on a cloth or paper mat will allow him to flip the finished puzzle over. If he has done it correctly, he should find a complete picture on the front!

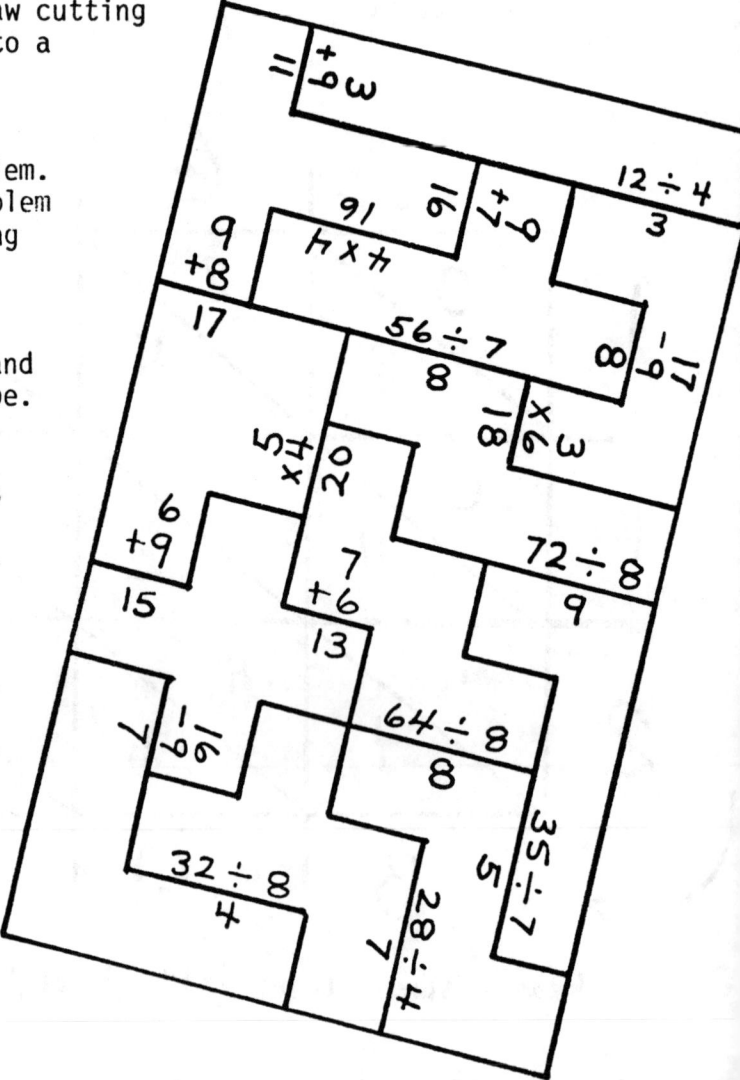

Variation: Students will want to make their own puzzles from psychedelic posters, magazine pictures, personal paintings, etc.

# FRACTIONAL NUMBERS

# STRING-A-FRACTION

**Skill:** *Using objects to show the value of fractions*

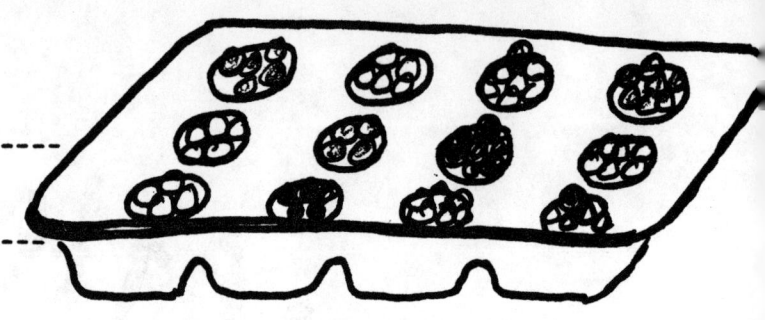

1. Fill each of the sections of a muffin tin with different colored wooden beads.

2. Place a set of fraction cards and several shoelaces with the muffin tin.

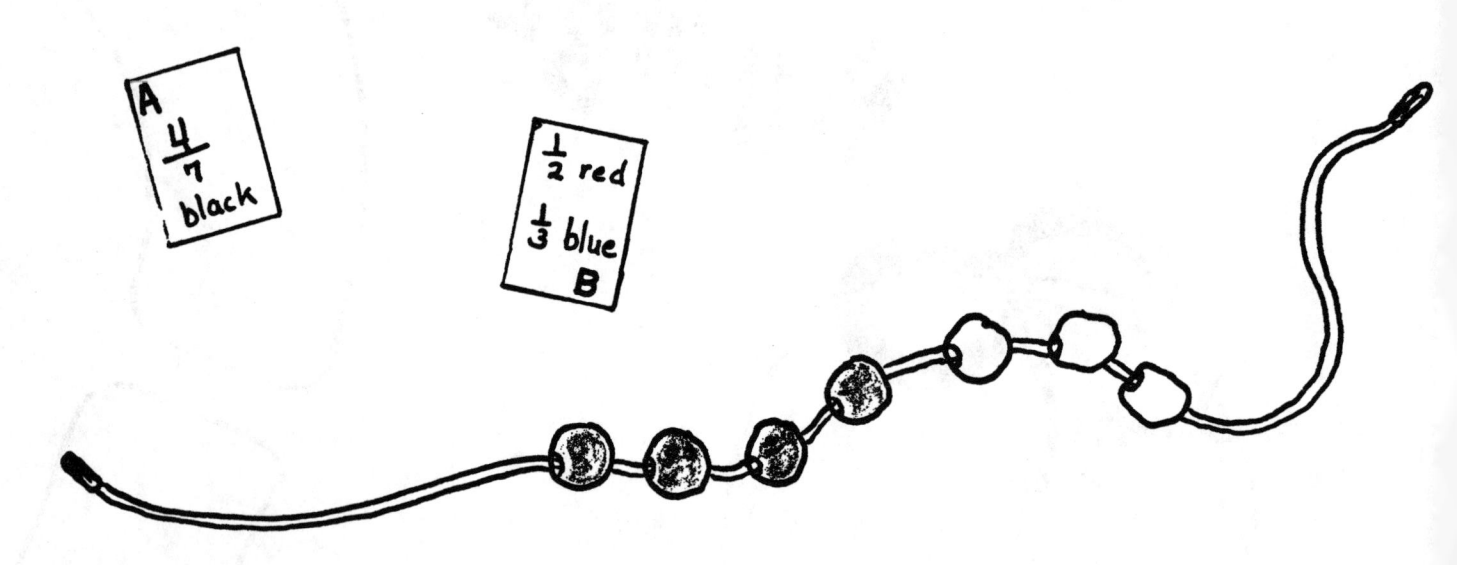

 A
 $\frac{4}{7}$
 black

 $\frac{1}{2}$ red
 $\frac{1}{3}$ blue
 B

3. Each card has a fraction and a color (Example A). Some cards may have two fractions and two colors (Example B).

4. Students draw a card and string a set that shows the fraction.

5. They may draw a picture of what they've made and label it with the fraction. In this way you can check their understanding of fractions.

**Variation:** Younger students will enjoy making and wearing fraction bracelets by "stringing" the beads on pipe cleaners.

FRACTIONS ON FILE
------------------------------------------------------------

Skill: *Using fractions and mixed numerals to name sets and parts of sets*

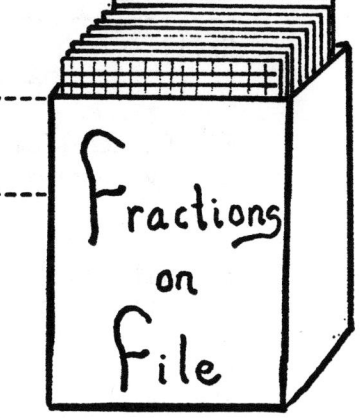

------------------------------------------------------------

1. Draw sets and parts of sets on graph paper. (See the samples below.)

2. Make several sheets of varying levels of difficulty.

3. With each new kind of question, provide a sample "How-to-Do-It" page.

4. Cover the pages with plastic sheet protectors and file them in order.

5. Provide an answer sheet so that students may check their own work.

6. The student takes a sheet, answers the questions by writing a fraction, checks his answers, and proceeds to a more difficult page.

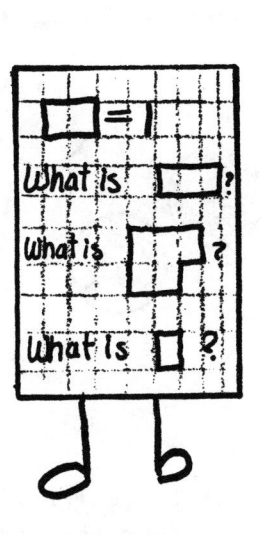

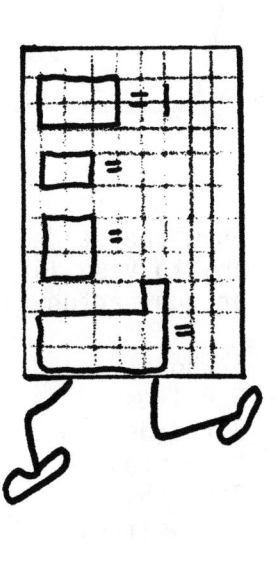

110

## CARTON CALCULATORS

---
Skill: *Adding and subtracting fractions with like denominators*
---

1. Change some ordinary egg cartons into addition-subtraction devices by decorating the outsides and pasting directions inside the cover.

2. With each carton include a plastic-covered problem sheet and a small bag of colored beads, beans, or marbles.

3. An egg carton can be used to solve problems with denominators of 1, 2, 3, 4, 6, or 12. For addends or sums greater than 1, use two cartons.

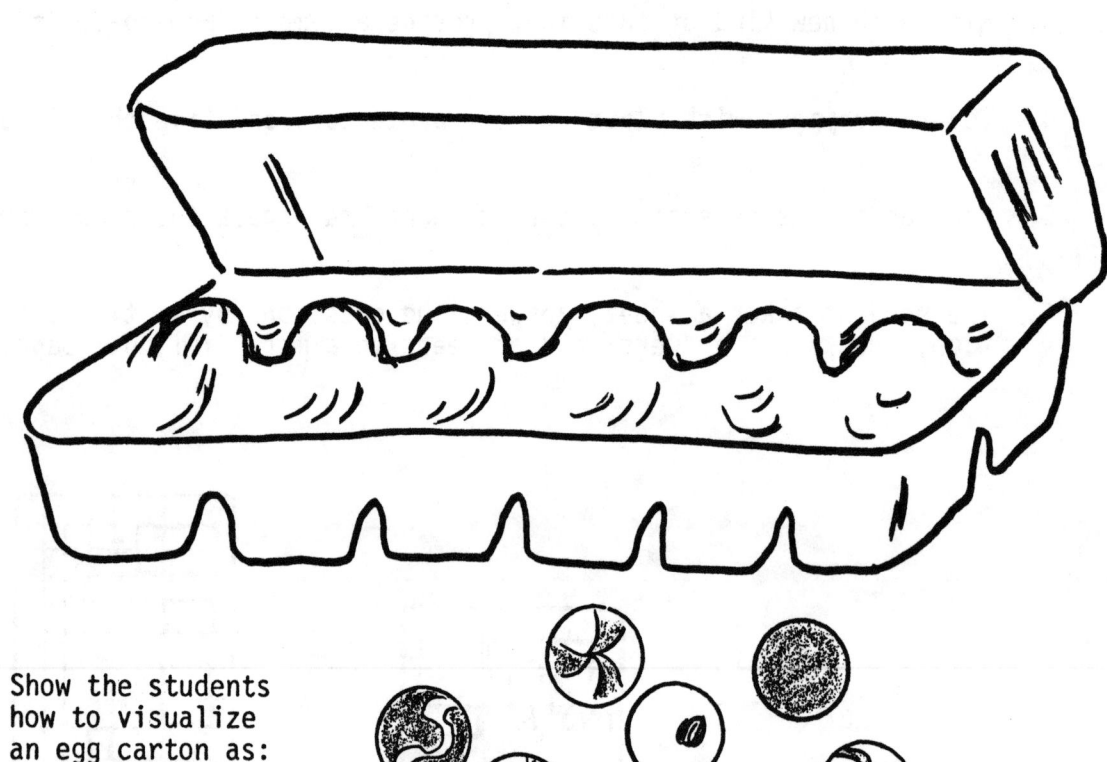

4. Show the students how to visualize an egg carton as:

    1 whole
    2 halves
    3 thirds
    4 fourths
    6 sixths
    12 twelfths

5. For each problem the student fills the number of sections representing the first fraction and either adds or subtracts the number represented by the second fraction, then counts what remains in the carton.

111

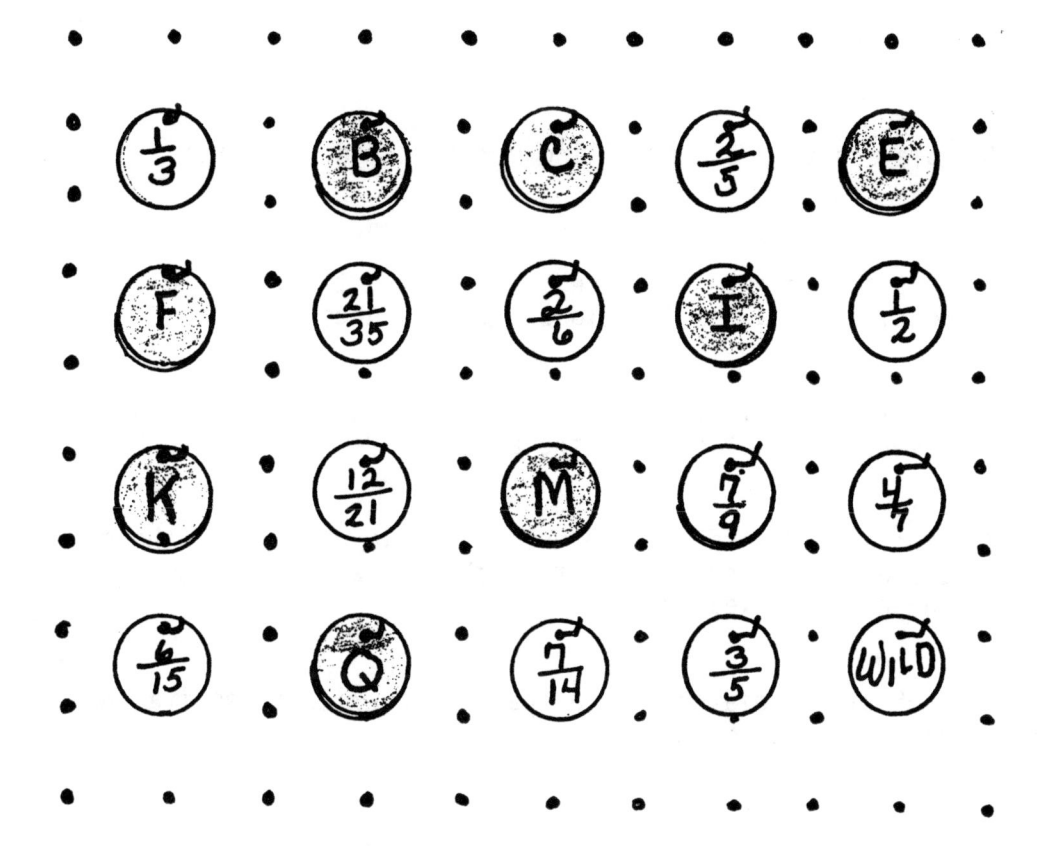

|   |   |   |   |   |
|---|---|---|---|---|
| $\frac{1}{3}$ | B | C | $\frac{2}{5}$ | E |
| F | $\frac{21}{35}$ | $\frac{5}{6}$ | I | $\frac{1}{2}$ |
| K | $\frac{12}{21}$ | M | $\frac{7}{9}$ | $\frac{4}{7}$ |
| $\frac{6}{15}$ | Q | $\frac{7}{14}$ | $\frac{3}{5}$ | WILD |

CONCENTRATION

------------------------------------------------------------
Skill: *Recognizing equivalent fractions*
------------------------------------------------------------

1. Make 48 cardboard circles (about 3 in. in diameter). Punch a hole in each circle.

2. On 24 circles, print letters A-X. On 22 others, write fractions, creating 11 pairs of equivalent fractions in all. Label the remaining two circles WILD.

3. Mix up the fractions (and WILD cards) and hang them in six rows on a pegboard or bulletin board. Hang the lettered circles in order <u>over</u> the fractions.

4. Two players take turns calling out a pair of letters:

   * If the pair reveals two equivalent fractions, the player continues and the fractions are left uncovered.

   * If two equivalent fractions do not show, the lettered covers are replaced and the other player takes a turn.

   * When a WILD card is uncovered, the player may win those spaces by naming any fraction that is equivalent to the one he has revealed.

   * Each time a player wins a "pair," he write an equivalency on his score sheet.

5. When the board is completed, players calculate how many pairs they have found. The winner is the one with the most pairs.

Variation: Make a third set of circles to hang behind the fractions. On these, write letters that will form a word, a sentence, a message, or an expression. Then, as each pair of equivalent fractions is located, part of the message will be revealed. Instead of playing for the most pairs, a player tries to decipher the message!

BEAU MATCH

---
Skill: *Recognizing equivalent fractions*
---

1. Display five to ten kites (with long tails) on a bulletin board.

2. On each kite write a fraction.

3. Make several bows for the kites' tails. On these, write fractions which are equivalent to those on the kites.

4. Place the bows in a central pocket and attach it to the board.

5. Make one large kite which explains how to match the fractions.

6. Hang another kite showing the cross product method for identifying equivalent fractions.

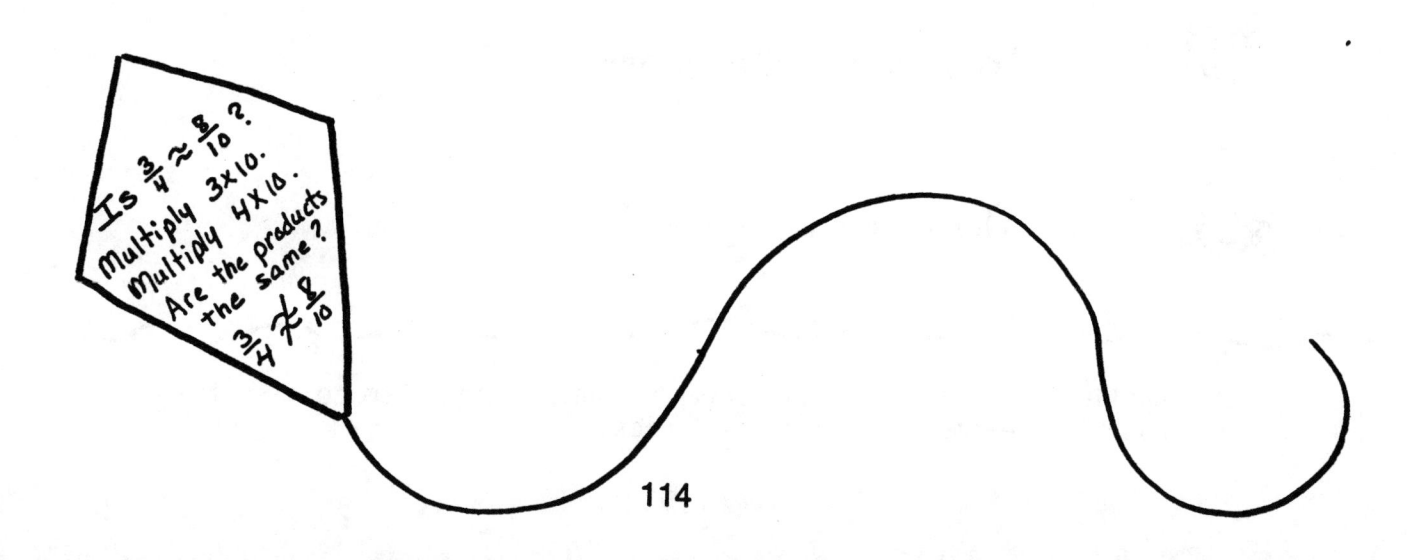

THINK BUNK!

---
Skill: *Writing sets of equivalent fractions*
---

1. To help students learn how to rename a fractional number with a new fraction, make a large Bunkbed Chart with these hints:

Think of a fraction as a pair of kids in a bunkbed who want to be treated equally!

Whatever one gets, the other one wants!

$\frac{2}{3}$ — Multiply me by 2    2 × 2 = 4

— Me too!    2 × 3 = 6

— Now you've got a new fraction $\frac{4}{6}$. It means the same amount as $\frac{2}{3}$. So, $\frac{2}{3}$ is equivalent to $\frac{4}{6}$.  ($\frac{2}{3} \approx \frac{4}{6}$)

$\frac{2}{3}$ — This time multiply me by 3.

— Don't forget me!

2. Duplicate for students an activity sheet asking them to give the bunkbed treatment to some other kids.

COMMON CUPBOARD

---
Skill: *Finding the greatest common factor of two numbers*

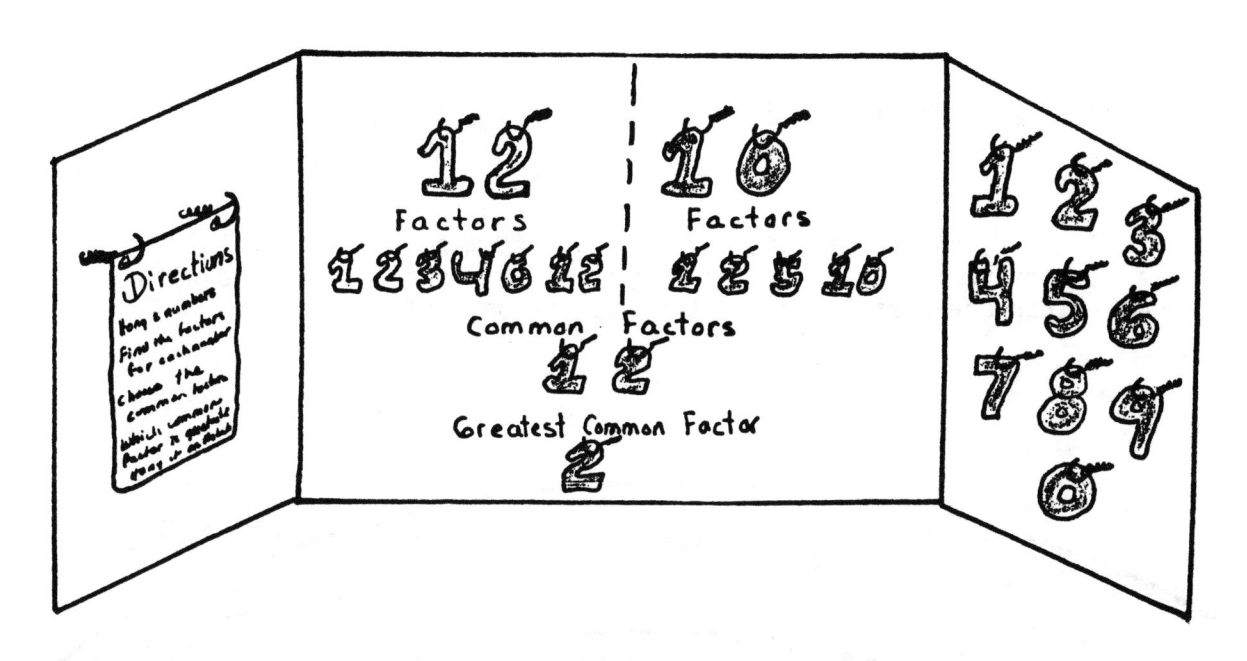

---

1. Cut away the top, bottoms, and one side of a thick cardboard box.

2. Have students paint or decorate both sides of the remaining piece.

3. Attach ten screw-type cup hooks to the right-hand section of the board.

4. On the hooks, hang the numerals 0-9 (four of each numeral).

5. Hang directions on the left-hand side of the board.

6. In the center, attach hooks as shown.

7. A student hangs two numbers, then finds the factors for each number. Then, he examines the factors to determine which factors are common to both numbers and which is the greatest common factor.

   Variation: This board might be used first by the teacher to introduce the idea of common factors to a group.

# FOLLOW THE FRACTION PATH

**Skill:** *Identifying fractions that are written in lowest terms*

1. Make a maze that is filled with fractions.

2. Tell the students that, to get through the maze, they must follow a path of fractions in lowest terms. (You might make a maze with more than one possible path, such as the one below.)

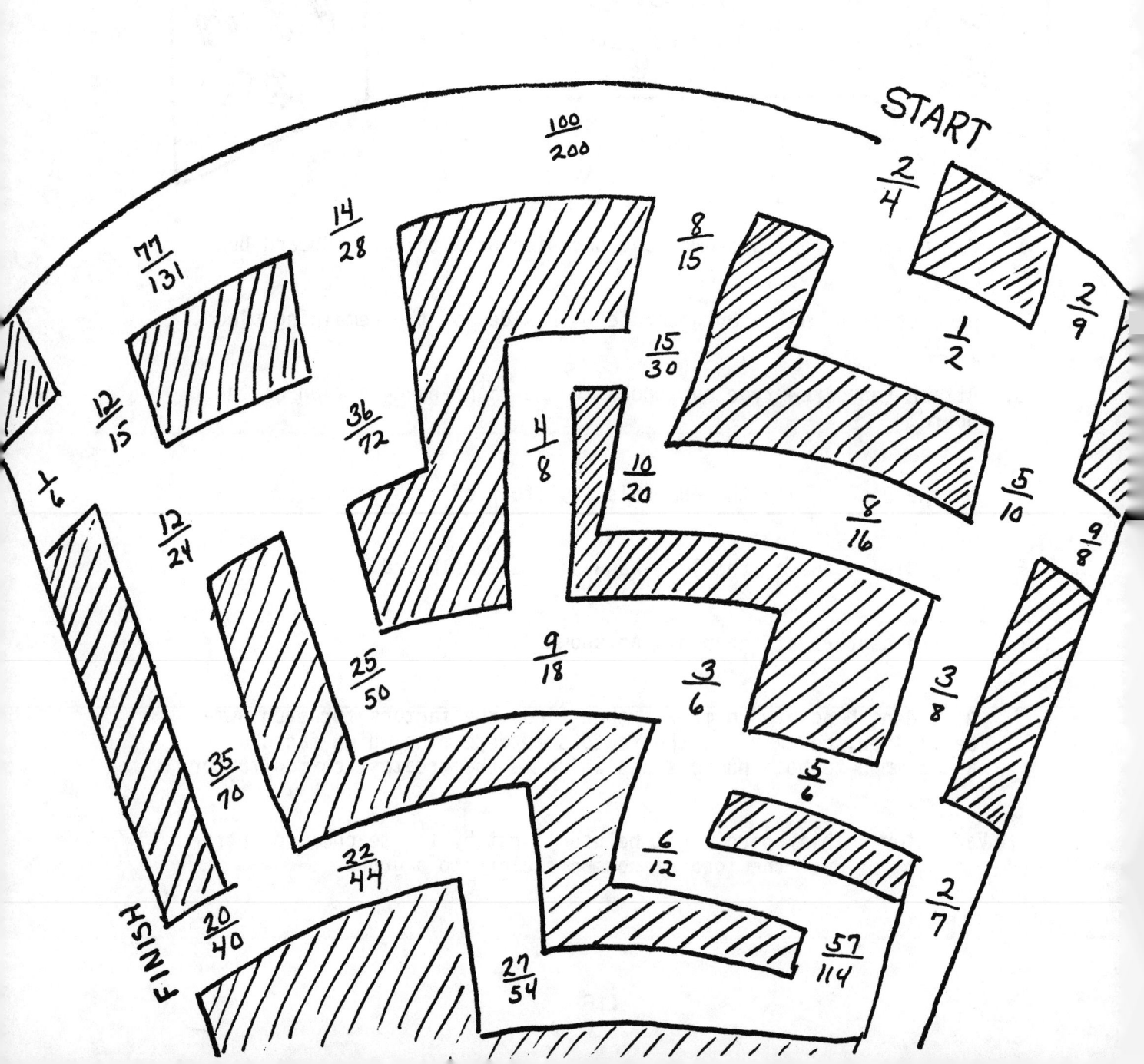

# CATERPILLARS TURN INTO BUTTERFLIES, YOU KNOW!

---
Skill: *Renaming unlike fractions as like fractions*
---

1. Fill a bag with caterpillars on which you have written fractions.

2. Make butterflies for another bag. Leave them blank and cover them with clear adhesive paper so they can be written on with crayon.

3. Ask a student to take two caterpillars and change the caterpillars into two butterflies with like denominators.

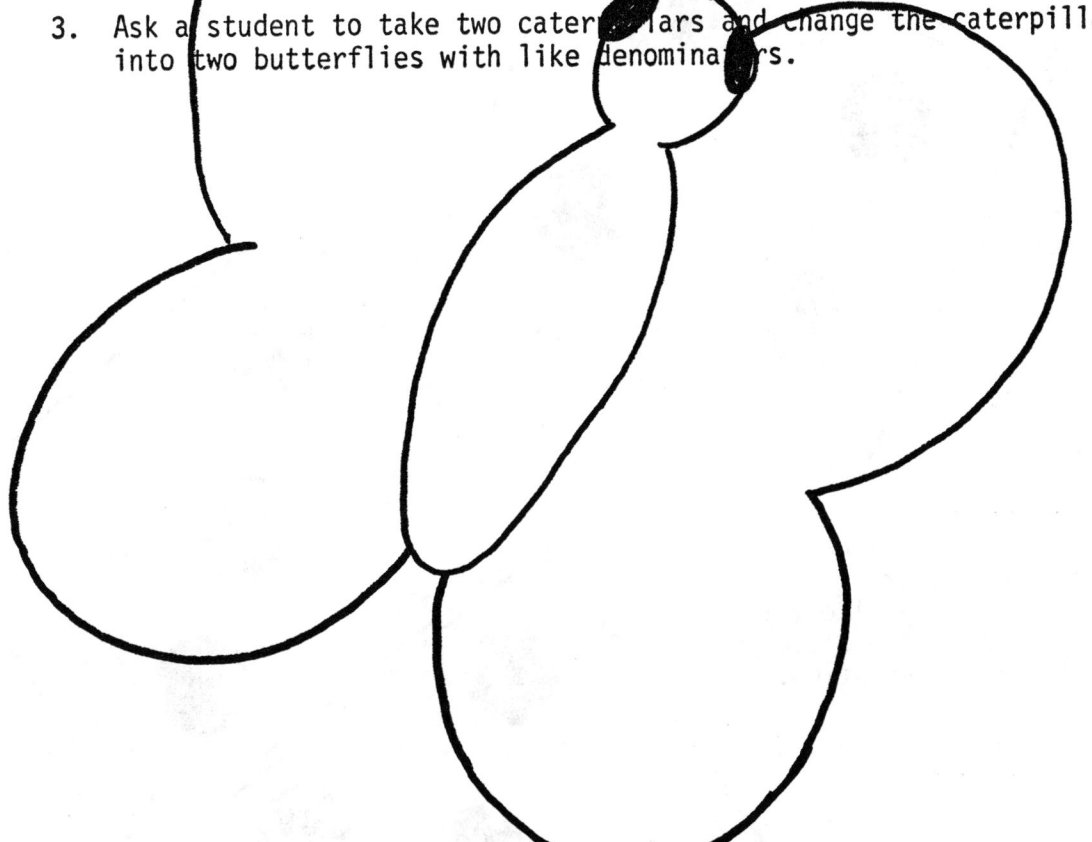

Variation: Make this into an addition exercise by adding the butterflies after they have been named. Call the activity YOU CAN'T ADD CATERPILLARS!

$\frac{4}{7}$  $\frac{2}{3}$ → $\frac{12}{21}$  $\frac{14}{21}$

## DOMIN-KNOWS!

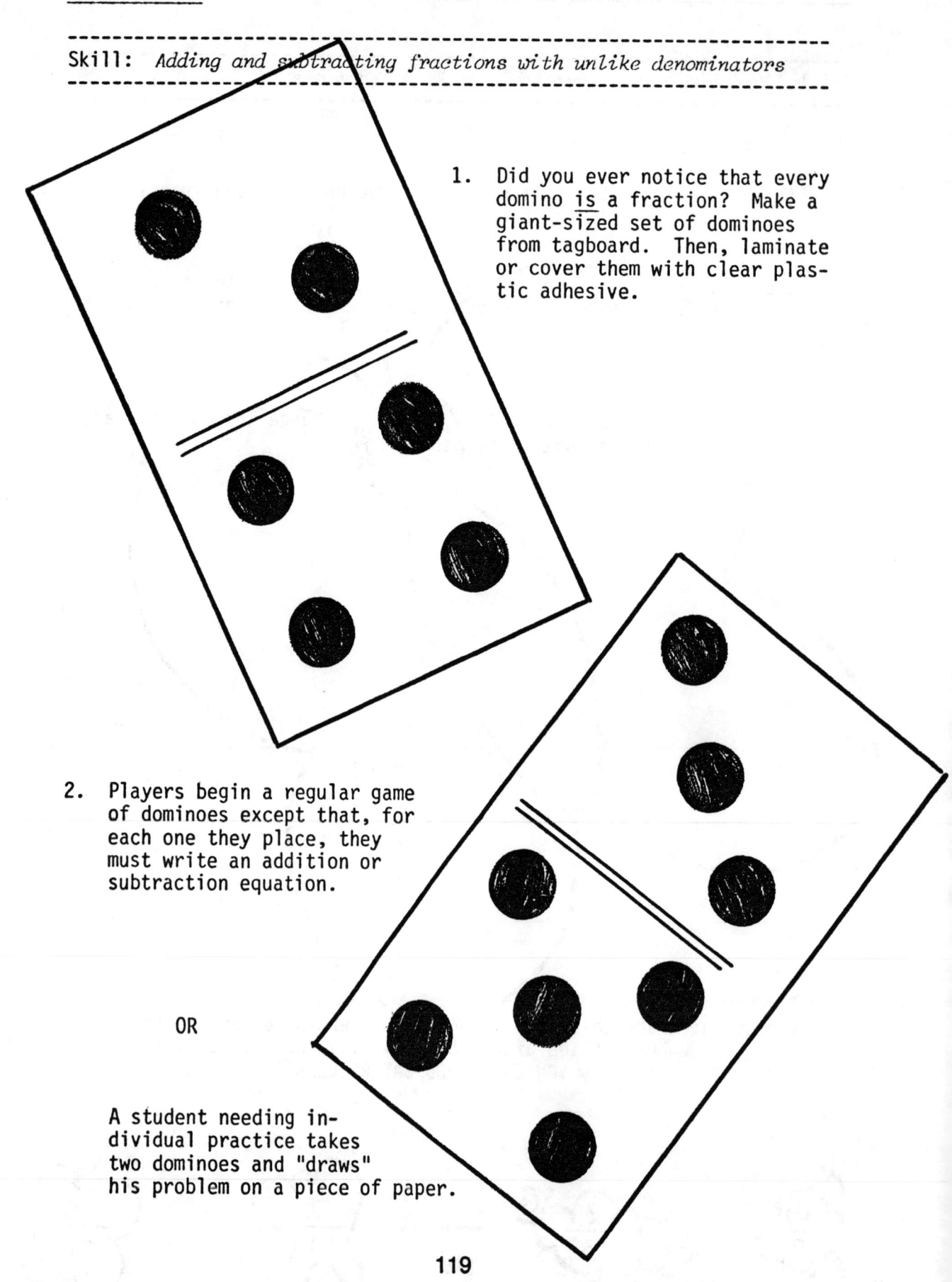

**Skill:** *Adding and subtracting fractions with unlike denominators*

1. Did you ever notice that every domino <u>is</u> a fraction? Make a giant-sized set of dominoes from tagboard. Then, laminate or cover them with clear plastic adhesive.

2. Players begin a regular game of dominoes except that, for each one they place, they must write an addition or subtraction equation.

   OR

   A student needing individual practice takes two dominoes and "draws" his problem on a piece of paper.

# BALANCE A BARBELL

**Skill:** *Expressing improper fractions as mixed numerals and vice versa*

1. Draw a group of weight-lifters showing off their muscles. Give each student a copy. (The illustration on this page may be used as a pattern.)

2. Write a mixed numeral or an improper fraction on one end of each barbell.

3. Direct the students to fill in the empty end with the correct mixed numeral or improper fraction.

$3\frac{1}{3}$ —— $\frac{10}{3}$     ? —— $\frac{17}{6}$

## ON YOUR MARK

---
**Skill:** *Adding and subtracting mixed numerals*
---

1. Set up a place in the room (or outside) where you can perform and measure the standing and/or running broad jump.

2. As each student jumps, measure the distance in feet or meters. Keep a list of each jump. Write the information in whole or mixed numerals.

3. For extra interest or to diminish the competition in the jumping, add other people (teachers, etc.) or try unusual jumps such as hopping on one foot or jumping with your feet tied together.

4. Make a series of problems requiring students to add and subtract the information they have collected.

# FRACTION FROLIC

---
Skill: *Ordering fractions by size*
---

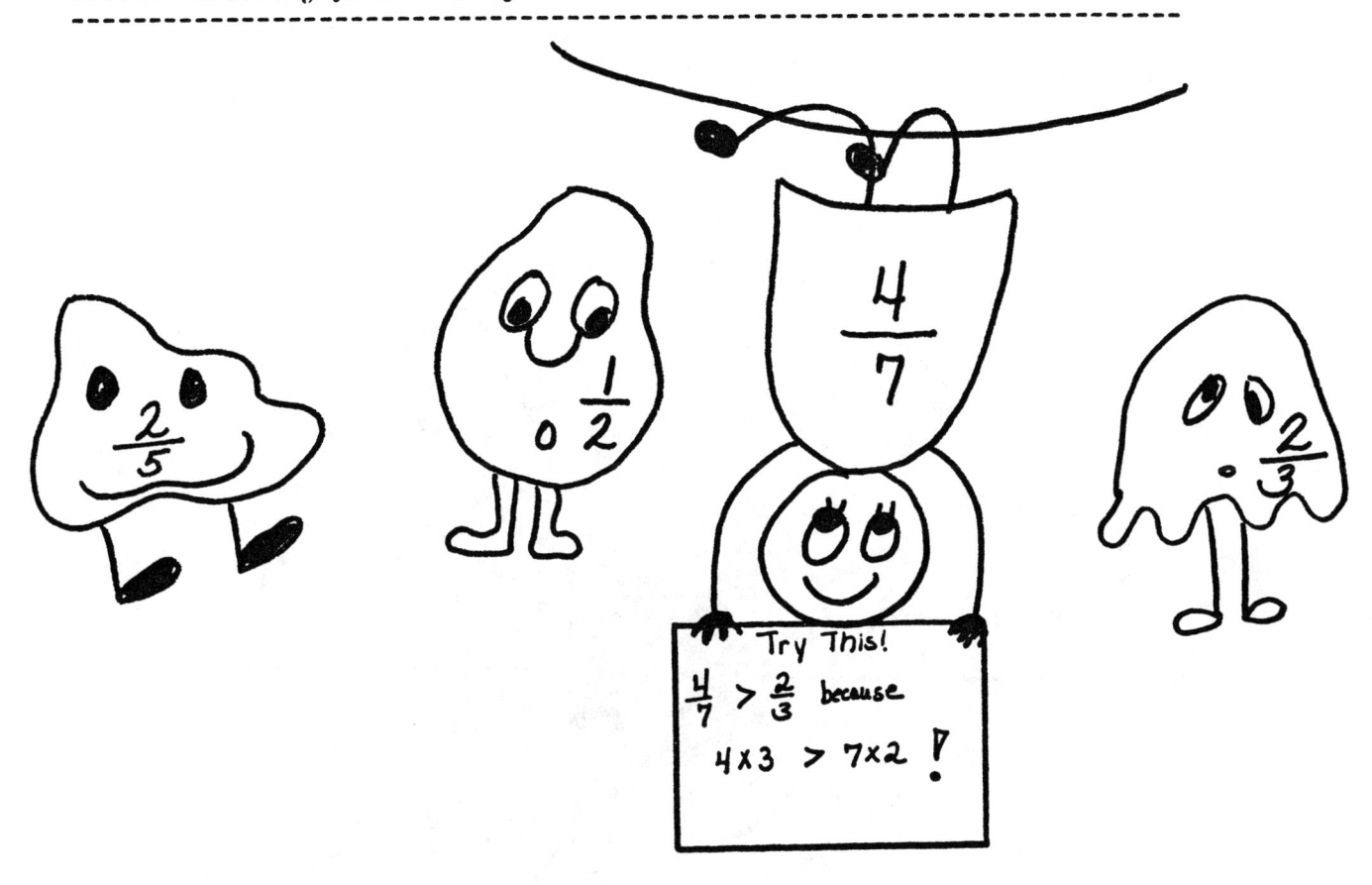

*Try This!*
$\frac{4}{7} > \frac{2}{3}$ because
$4 \times 3 > 7 \times 2$ !

1. Provide a set of about 25 fractions to be put in order. Pin one to each student.

2. Show students this method for ordering fractions:

> To compare fractions, cross multiply.
> $\frac{4}{5} \:\: \frac{2}{3}$
> If product ① is greater, the first fraction is greater.
> If product ② is greater, the second fraction is greater.

3. Divide the group into two or three teams and let them order themselves.

# EASY BULL'S-EYE

---
Skill: *Multiplying fractions*
---

1. Make a multi-colored target about three feet in diameter.

2. Write fractions on the target.

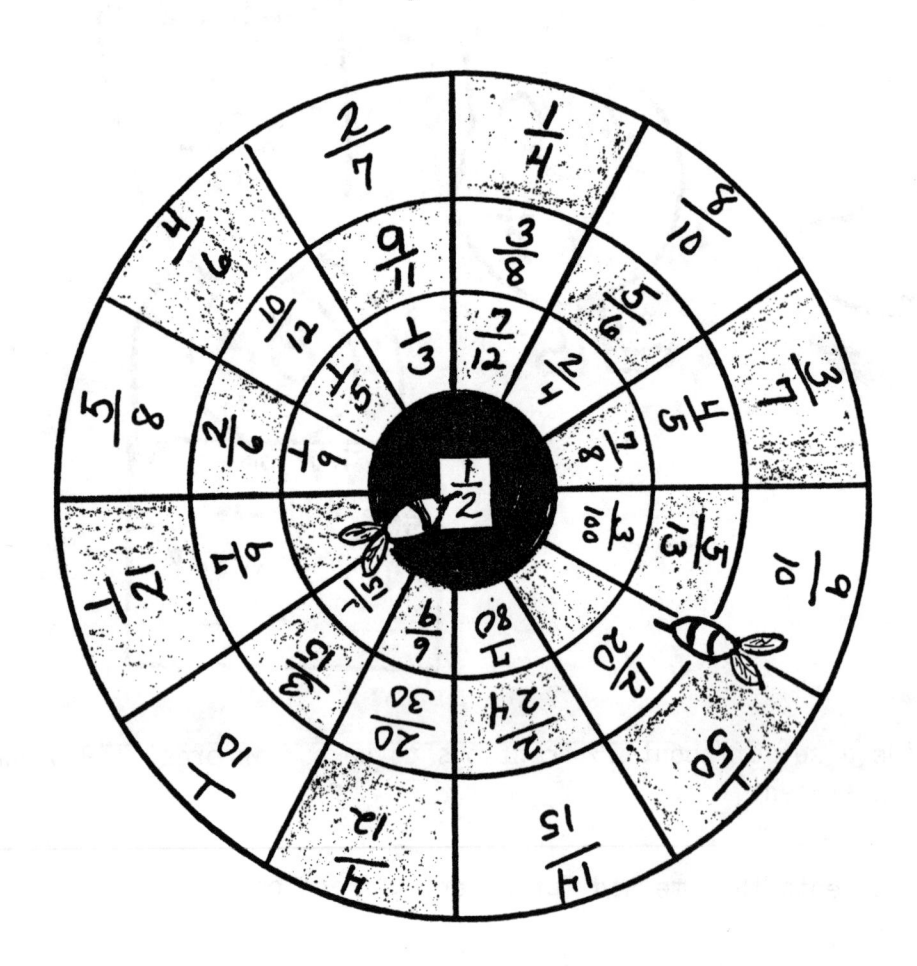

3. For multiplication practice, a student throws two darts and writes a multiplication equation.

4. His score will depend on the accuracy of his answers.

   Variation: Let students make up rules for a game in which two or more players compete. The products may be given orally, and the players check one another by figuring out the answer on scrap paper.

BOIL, BAKE, OR FRY A FRACTION!

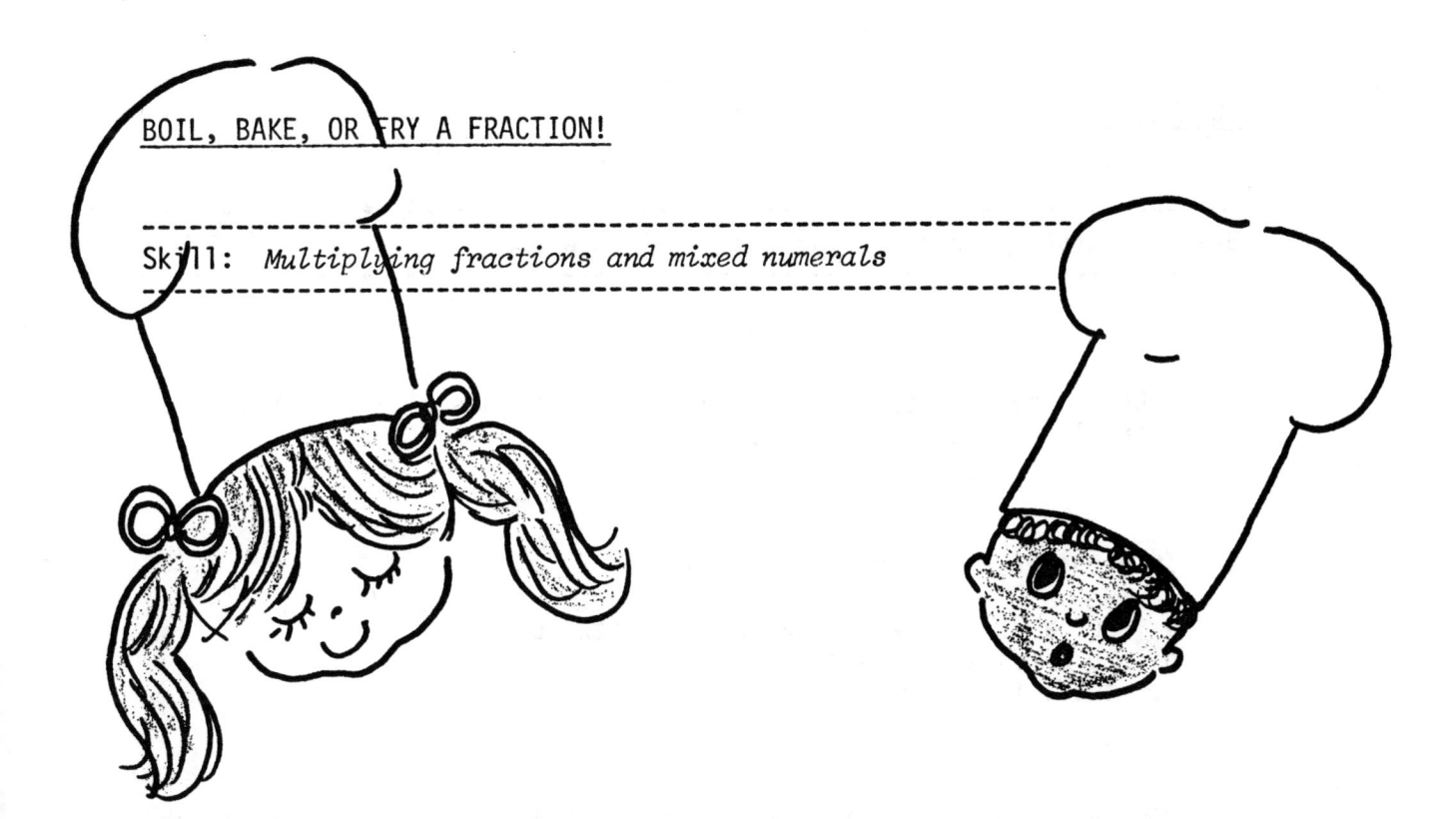

Skill: *Multiplying fractions and mixed numerals*

1. Set up a cooking center where students can make simple "dishes" (fudge, marshmallow-cereal candy, jello, pudding, etc.).

2. Ask students to contribute recipes. Try to include many recipes with fractions.

3. Make Task Cards that require students to multiply fractions.

4. Assign Task Cards to individuals or small groups. Some tasks may be for practice in changing recipes and, therefore, may not necessarily involve cooking.

5. Accurate completion of these activities will be rewarding for all!

*Write a recipe for $\frac{1}{3}$ batch of fudge*

*Make $2\frac{1}{2}$ batches of cherry jello.*

# EQUATION RACE

---
Skill: *Multiplying fractions and mixed numerals*
---

1. Make two sets of cards approximately six inches square. Cover them with plastic adhesive for durability.

2. In each set, include:

    a. 20 cards with whole numbers 0-9 (two of each)
    b. 20 cards with fractions
    c. two symbol cards (x and =)

3. Divide students into two teams. Put all the cards in two piles on the floor.

4. Dictate two factors. Each team must use its set of cards to write the problem and show the answer.

5. Give one point to the team that correctly completes the problem first.

6. Begin a new round with another pair of fractions and/or mixed numerals.

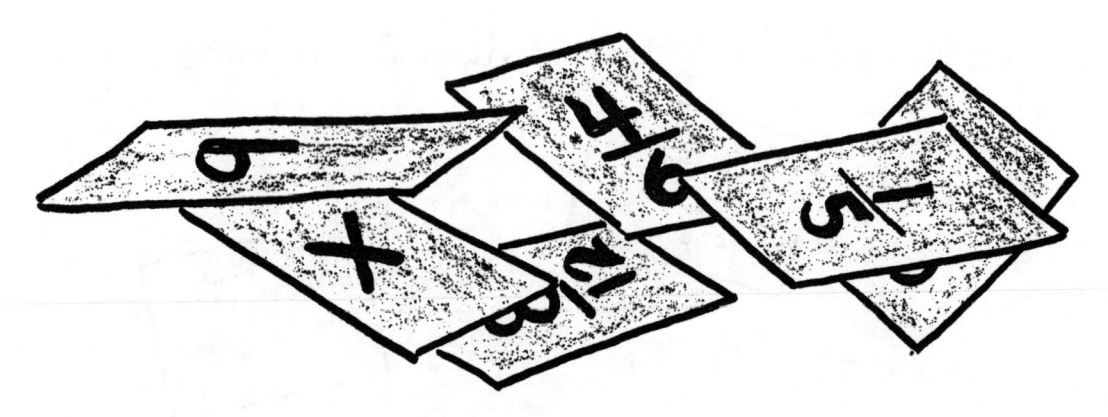

## UPSIDE DOWN MAKES IT RIGHT

Skill: *Dividing fractions using the reciprocal method*

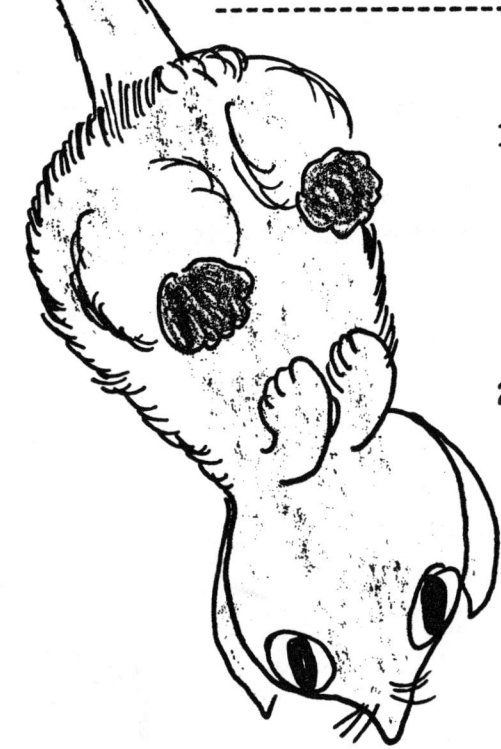

1. Give students worksheets similar to the two shown here. Provide glue, scissors, pencils, and crayons.

2. In each problem on Sheet A, they must:

   a. locate the divisor

   b. find the reciprocal of the divisor on the Sheet B

   c. cut out that reciprocal

   d. paste it over the divisor

   e. solve the problem by multiplying

| A | B |
|---|---|
| $\frac{7}{8} \div \frac{2}{3} =$ | $\frac{3}{1}$   $\frac{2}{1}$ |
| $\frac{1}{5} \div \frac{7}{8} =$ | $\frac{7}{3}$   $\frac{5}{2}$ |
| $\frac{2}{9} \div \frac{1}{3} =$ |  |
| $\frac{6}{10} \div \frac{2}{5} =$ | $\frac{8}{7}$   $\frac{4}{5}$ |
| $\frac{9}{12} \div \frac{1}{2} =$ |  |
| $\frac{8}{17} \div \frac{3}{7} =$ | $\frac{9}{10}$   $\frac{3}{2}$ |
| $\frac{6}{16} \div \frac{5}{4} =$ |  |

WRITE-A-RATIO

Skill: *Using a fraction to name a ratio*

1. Cut out a variety of pictures representing different sets.

2. Mount two sets on each of several pieces of construction paper.

3. Show one such pair to students. Ask them to give two different ratios to compare the two quantities shown in the pictures.

4. For the remaining pairs, students work individually to write two ratios for each sheet.

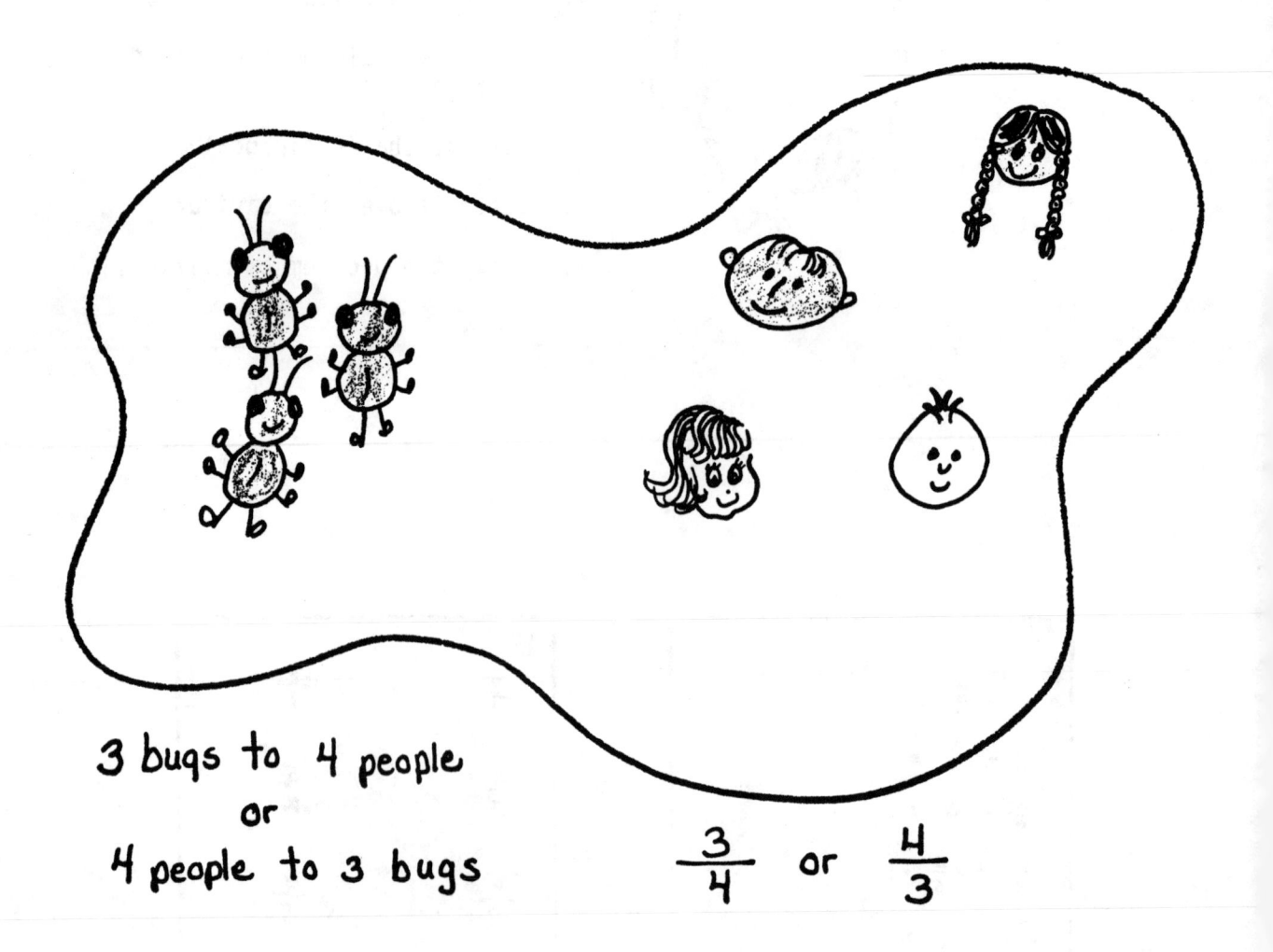

3 bugs to 4 people
or
4 people to 3 bugs

$\frac{3}{4}$ or $\frac{4}{3}$

# THINK DELICIOUS!

---
Skill: *Writing and solving proportions*
---

1. Find four large glass jars with wide mouths.

2. Label the lids as follows:

    1--NUMBERS
    2--STORIES TO SOLVE
    3--I'M DONE!
    4--DELICIOUS!

3. Cut "cookies" from brown construction paper.

4. In Jar 1 put "cookies" with whole numbers.
   In Jar 2 put "cookies" with short story problems involving proportions.

5. Write directions on the outsides of the jars. Include two or three levels of directions on each jar. From Jar 1, students take three number cookies and arrange them to make a proportion. They fill in the fourth number.

   From Jar 2, students take one cookie at a time and solve the story problems. Ask some students to write their own stories on the blank cookies.

   **COOKIES**

6. Near the jars, place a sign explaining the requirements for each level at the center.

7. Assign a particular level to a student. He will read and follow the directions on the jars, writing his answers on paper. His finished paper will go into Jar 3 for checking.

8. Fill Jar 4 with real cookies. As a student places his completed paper in Jar 3, he may take a cookie from Jar 4!

    DELICIOUS!

PIN-ON-PROBOSCIS

---
Skill: *Matching ratios with percentages*
---

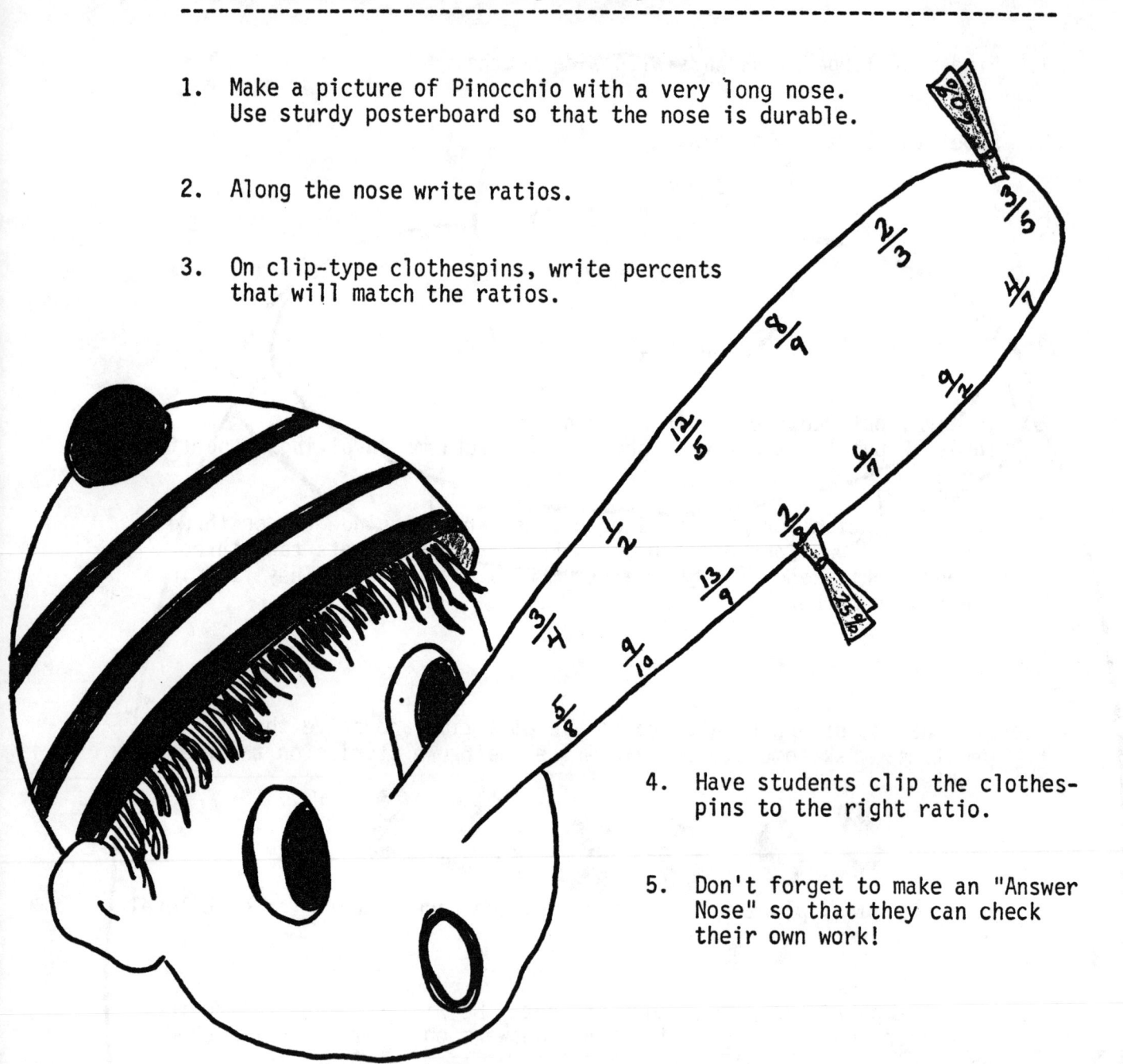

1. Make a picture of Pinocchio with a very long nose. Use sturdy posterboard so that the nose is durable.

2. Along the nose write ratios.

3. On clip-type clothespins, write percents that will match the ratios.

4. Have students clip the clothespins to the right ratio.

5. Don't forget to make an "Answer Nose" so that they can check their own work!

Alternative: Give Pinocchio a new nose and students can match fractions with decimals or mixed numerals with decimals.

129

PER-SENSE

---
Skill: *Finding the percent of a number*
---

1. Make fifty or more activity cards that will give practice finding percentages.

    Suggestions:

    a. Make a personal budget. What percent of your earnings do you save?

    b. What percent of the cars in the school parking lot are American-made?

    c. What percent of the students in the school are absent today?

    d. What percent of your classmates have freckles?

    e. Time someone swinging at recess. What percent of the recess does he stay on the swing?

    f. What percent of your class brought lunch today?

    g. What percent of the boys have white gym shoes?

    h. What percent of your day do you spend sleeping? eating? playing? at school?

2. Put these into a file box, dividing them into sections according to their difficulty.

3. Students may work independently at the tasks they choose.

4. It might be helpful if a reminder about "How to Find Percents" could be taped inside the lid!

DIAL-A-DECIMAL DUO

---
Skill: *Using decimals in addition, subtraction, multiplication and division problems*
---

1. Make some phones for dialing decimal pairs. Prepare three phones with decimals at three levels of difficulty.

2. A student "dials" two decimal numerals and uses the two in four problems (one for each operation).

   Variations:  1. Change the numerals on the dial often.

   2. This activity might be limited to one or two operations, if students are not yet studying multiplication and division of decimals.

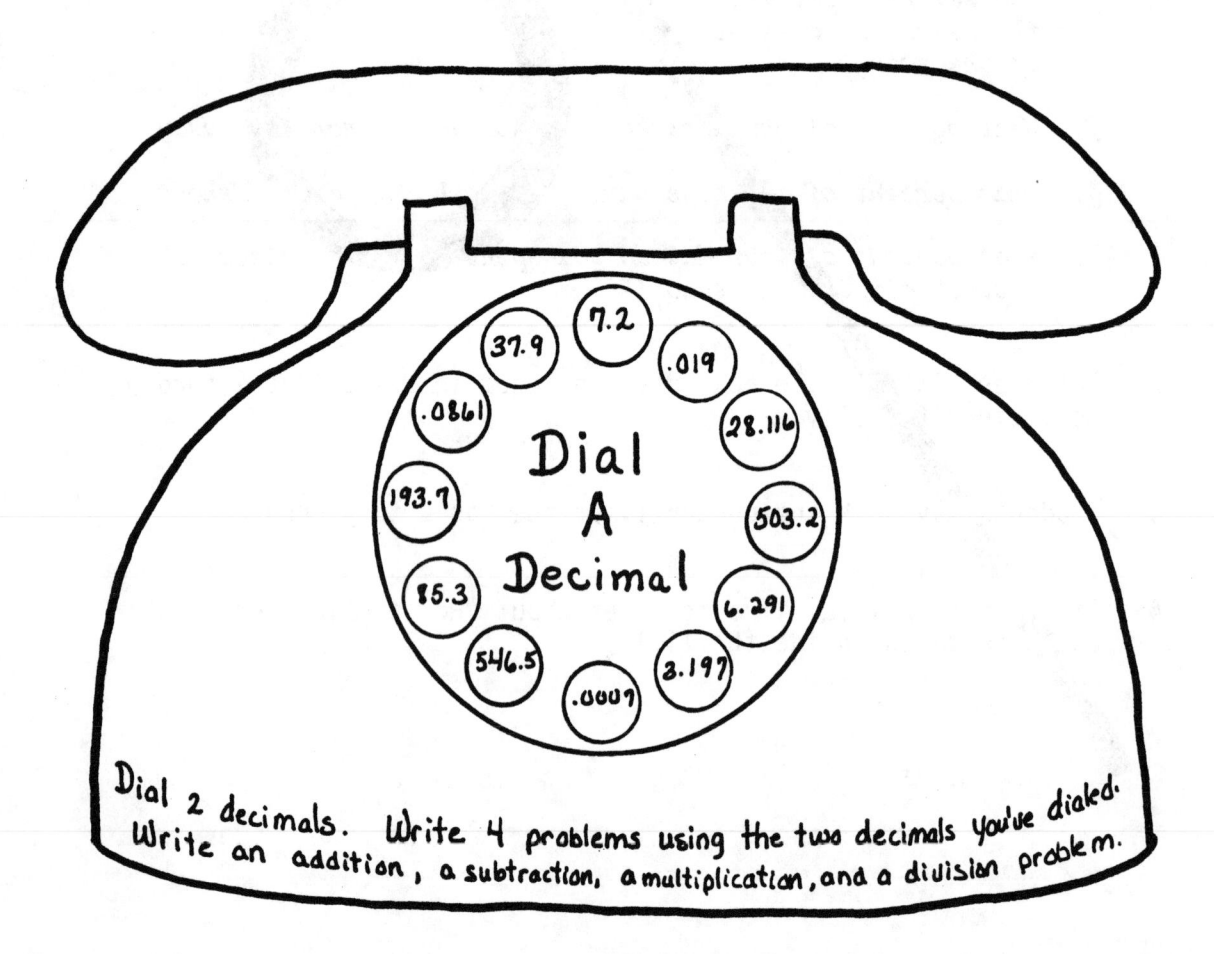

*Dial 2 decimals. Write 4 problems using the two decimals you've dialed. Write an addition, a subtraction, a multiplication, and a division problem.*

Dial A Decimal numbers: 7.2, 37.9, .019, .0861, 28.116, 193.7, 503.2, 85.3, 6.291, 546.5, .0009, 2.197

# PROBLEM SOLVING

CENTER ON SOLUTION

---
Skill: *Creating and solving problems through activities using manipulatives, pictures, and dramatization*
---

1. Set up a do-it-yourself Drama Math Center in which you provide lots of materials for creating and demonstrating problems and their solutions.

2. In the center, include:

    --several colored pictures that suggest problems

    --an assortment of objects to manipulate (shoes, beads, straws, glasses, blocks, play money, hats, boxes, cookies, etc.)

    --a box of old clothes and disguises which would encourage role-playing (hats, wigs, skirts, cane, moustache, etc.)

    --a box of cards containing problems to be solved

3. Introduce the center by creating a problem or stating one verbally and choosing several students to find and show the solution.

4. Give each student a problem card and provide time for him to decide on his procedure for solving it.

5. Let individuals or groups demonstrate their problems before the class. Later, assign pairs of students to work in the center.

*What percent of the people are getting wet?*

| Students at Cherokee School Who've Had Chicken Pox | |
|---|---|
| Grade 1 | 🧍🧍🧍🧍🧍🧍🧍 |
| Grade 2 | 🧍🧍🧍🧍🧍🧍 |
| Grade 3 | 🧍🧍🧍🧍🧍🧍🧍🧍 |
| Grade 4 | 🧍🧍🧍🧍🧍🧍🧍🧍🧍 |
| Grade 5 | 🧍🧍🧍🧍🧍🧍🧍🧍🧍🧍🧍🧍 |
| Grade 6 | 🧍🧍🧍🧍🧍🧍🧍🧍🧍 |

🧍 = 3 students

## PICTURE THAT

---
Skill: *Answering questions using information in picture graphs*
---

1. Have students gather information pertaining to your school. They may survey and collect statistics on such things as:

    a. number of students in each class
    b. number of students with freckles in each grade
    c. favorite TV programs of the school population, etc.

2. Make picturegraphs to show the results of the surveys. To each graph, attach an envelope containing questions about the information pictured.

3. Students may answer the questions orally or write them down.

4. As a follow-up activity, students may create original graphs using the information gathered.

# SMARTY CAT CUBBY

**Skill:** *Solving problems requiring manipulation of objects*

*Come in?! Try to outsmart me*

SMARTY CAT

1. Start a collection of puzzles and brain-teasing problems which students must manipulate in order to solve. (Many math texts have examples of such sprinkled throughout, and there are dozens of puzzlers available commercially).

2. Create a Smarty Cat Cubby by placing a desk or table inside a large box in which puzzles can be stored. The sides of the box are handy spots for hints, suggestions and directions for making up new puzzles.

3. Encourage students to create or borrow brain teasers and add to the collection.

*bright idea!*

Some sources of problem solving opportunities

ice cream shop        mail order catalogs
newspaper             book orders
gas station           school lunch program
family budgets        grocery stores
school statistics     calories eaten during the

PROBLEM PIG

Skill: *Writing and solving word problems using given set of numbers*

1. Make a large pink piggy bank from poster board. Behind the slit, place an envelope containing "coins" on which numbers have been written. On some of the "coins" write amounts of money such as $37.00, 4¢, etc.

2. Near the bank place an envelope filled with small pigs which have been cut from construction paper.

3. A student chooses two (or more) "coins" from the bank. On a small pig he writes an original problem using those numbers.

4. Display the "problem piggies" that the students make along with the bank. These can be used later for more problem solving.

## BIB MATH

**Skill:** *Formulating original problems and equations*

1. Cut about forty bibs from tagboard. Attach a string to each so that it can be hung around the neck.

2. On the bibs write the digits 0-9 three times (using thirty bibs).

3. Use the other bibs for writing symbols:

    $+ \quad - \quad \times \quad \div \quad < \quad > \quad = \quad \approx \quad . \quad \neq \quad \$ \quad ¢$

4. When the students are wearing the bibs, ask them to create some problems. For example:

    *Show four addition equations.*
    *Make an equation which uses two operations.*
    *Make two money problems.*
    *Show an inequality between two fractions.*
    *Make a division equation with fractions.*
    *Show an equation using decimals.*
    *Show a multiplication problem with negative integers.*

    **Alternative:** Make more symbols so that students may work as two teams, each trying to complete the problems first.

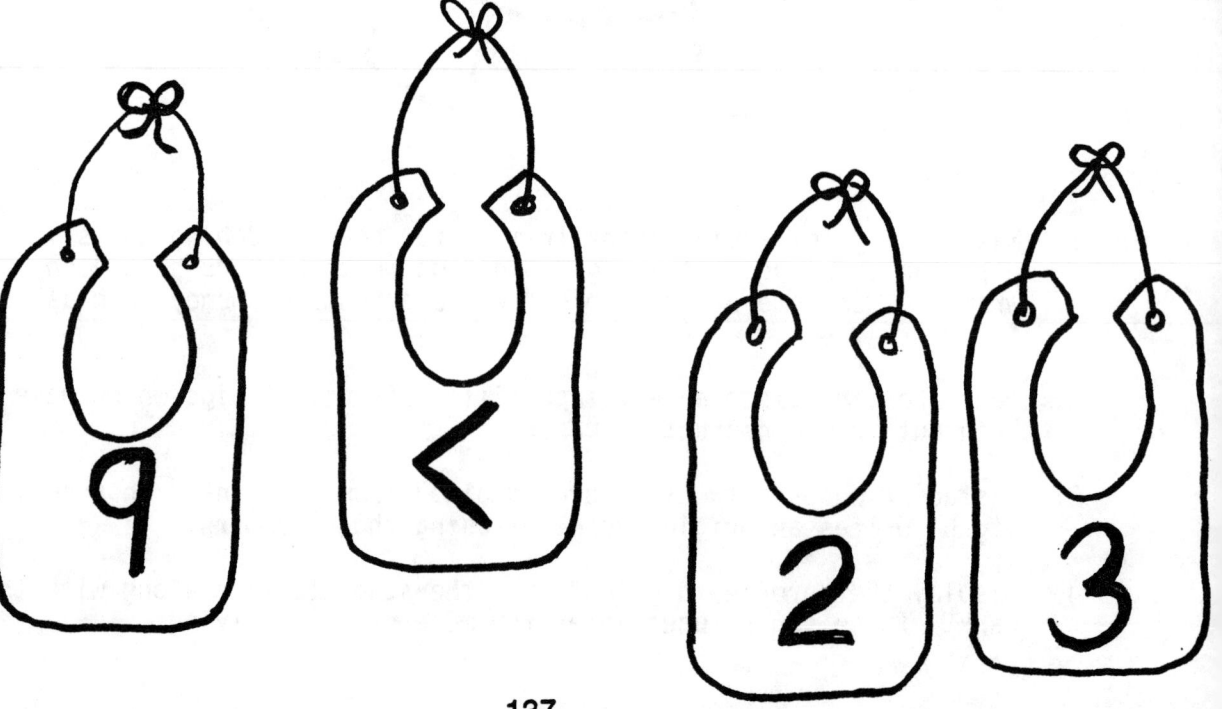

# SHOPPING SPREE

---
Skill: *Writing and solving equations from given information*
---

1. Set up a Shoppers' Research Station. Provide:

   adding machine (if available)
   pencils
   paper
   catalogs:  Sears catalog
              seed catalog
              discount store catalog
              toy catalog
              office equipment catalog
              book catalog
              camping goods catalog
              appliance catalog
              automotive parts catalog
              sporting goods catalog

   newspapers
   flyers from grocery and drug stores
   three small shopping bags

2. Using adding machine tape or 3 x 5 cards, write a series of tasks requiring students to locate items, compare prices, and solve problems.

3. Group the TASKS into three levels and place them in the appropriate shopping bag.

4. Have a pile of sheets on which students can compute and share their answers.

     OR

   Ask students to make a poster advertising the product they choose as the best buy.

     OR

   Show students how to make coupon books containing one coupon for each item they decide to buy!

## HEY, STACK!

---
**Skill:** *Formulating and solving original problems*
---

1. Write numerals and symbols on all six sides of several wooden cubes.

   (Foam cubes can be cut from large foam rubber cushions. Magic marker will write effectively on these. The foam cubes are light to carry and not very noisy!)

2. Ask students to stack an equation to meet various specifications. Those directions might be oral, or they may be written on blocks too.

3. Provide an egg timer or stopwatch for "timed stacking."

   OR

   Make two identical sets of blocks, and the activity can become a contest.

*How many equations can you stack in 3 minutes?*

*Use 4 blocks to write a multiplication sentence.*

*Write an addition equation.*

*Make a problem using all the blocks.*

# MAKE A FAST THOUSAND!

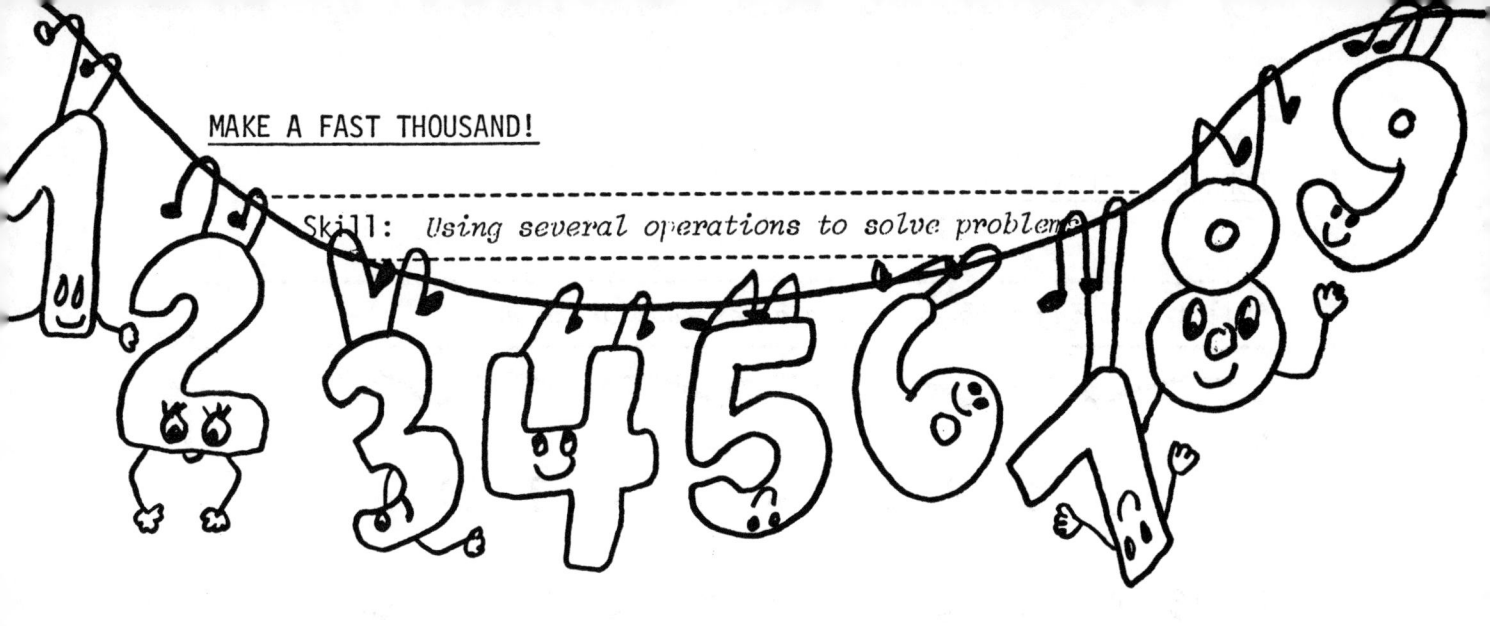

**Skill:** *Using several operations to solve problems*

1. Make nine square cards bearing the numerals 1-9. Cover them with plastic for durability.

2. Give students the following directions:

   *Use these nine digits to make 1000.*
   *Use all of them.*
   *You may arrange or combine them in any manner.*
   *You may use any operations.*
   *Write down the equation you create.*

   *P.S. There are several possible solutions! Try more than one!*

3. Students may work individually or in pairs to find a solution. A large piece of manila paper will provide a good work area because they can write on the paper between the numerals as necessary.

4. Keep a list of all the solutions discovered by the students.

**ght idea!**

Do lots of mental calendar and clock problems with your group, i.e.:

What day is two weeks and four days from today?
What time will it be in 6¼ hours?
What was the date seven weeks ago?

# CORNER TO CORNER

**Skill:** *Solving problems using a calendar or chart*

| | | | | | | | | | | | |
|---|---|---|---|---|---|---|---|---|---|---|---|
| 1 | 2 | 3 | 4 | 5 | 6 | 7 | 8 | 9 | 10 | 11 | 12 |
| 13 | 14 | 15 | 16 | 17 | 18 | 19 | 20 | 21 | 22 | 23 | 24 |
| 25 | 26 | 27 | 28 | 29 | 30 | 31 | 32 | 33 | 34 | 35 | 36 |
| 37 | 38 | 39 | 40 | 41 | 42 | 43 | 44 | 45 | 46 | 47 | 48 |
| 49 | 50 | 51 | 52 | 53 | 54 | 55 | 56 | 57 | 58 | 59 | 60 |
| 61 | 62 | 63 | 64 | 65 | 66 | 67 | 68 | 69 | 70 | 71 | 72 |
| 73 | 74 | 75 | 76 | 77 | 78 | 79 | 80 | 81 | 82 | 83 | 84 |

1. Display a large chart or calendar with numbers. Place it on the floor.

2. Make 40-50 cards with symbols such as the following:

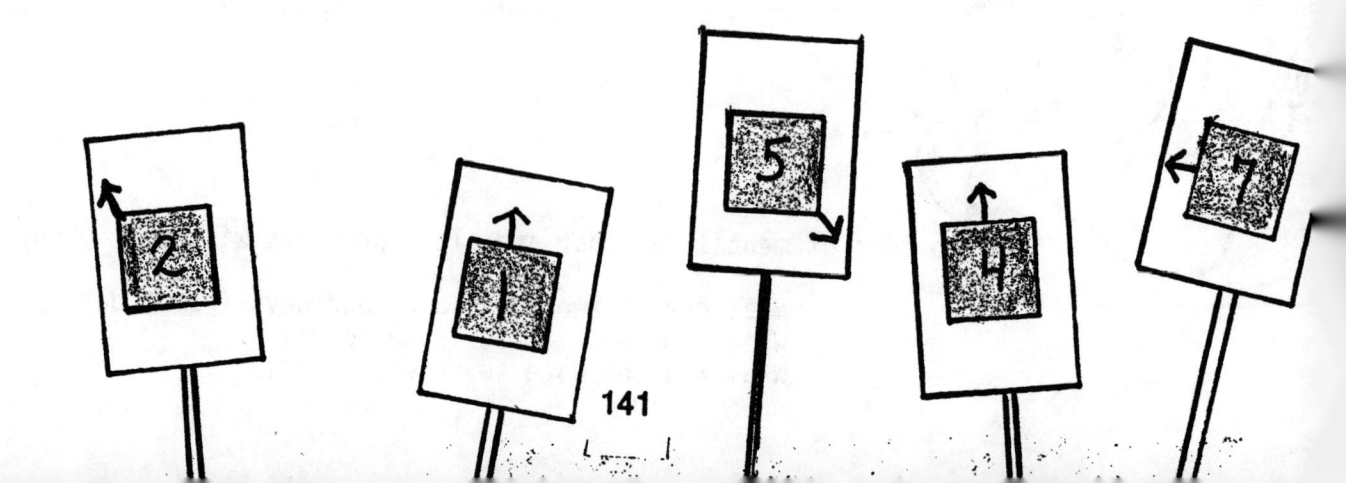

3. Provide four stiff cardboard squares of various colors. These will be the movable markers in the game. Two to four players each place their markers in a corner of the chart.

4. Each draws a card in turn and moves in accordance with the symbol. The first player to reach the corner opposite his starting place is the winner.

5. Later on, you may ask the players to write an equation for each move.

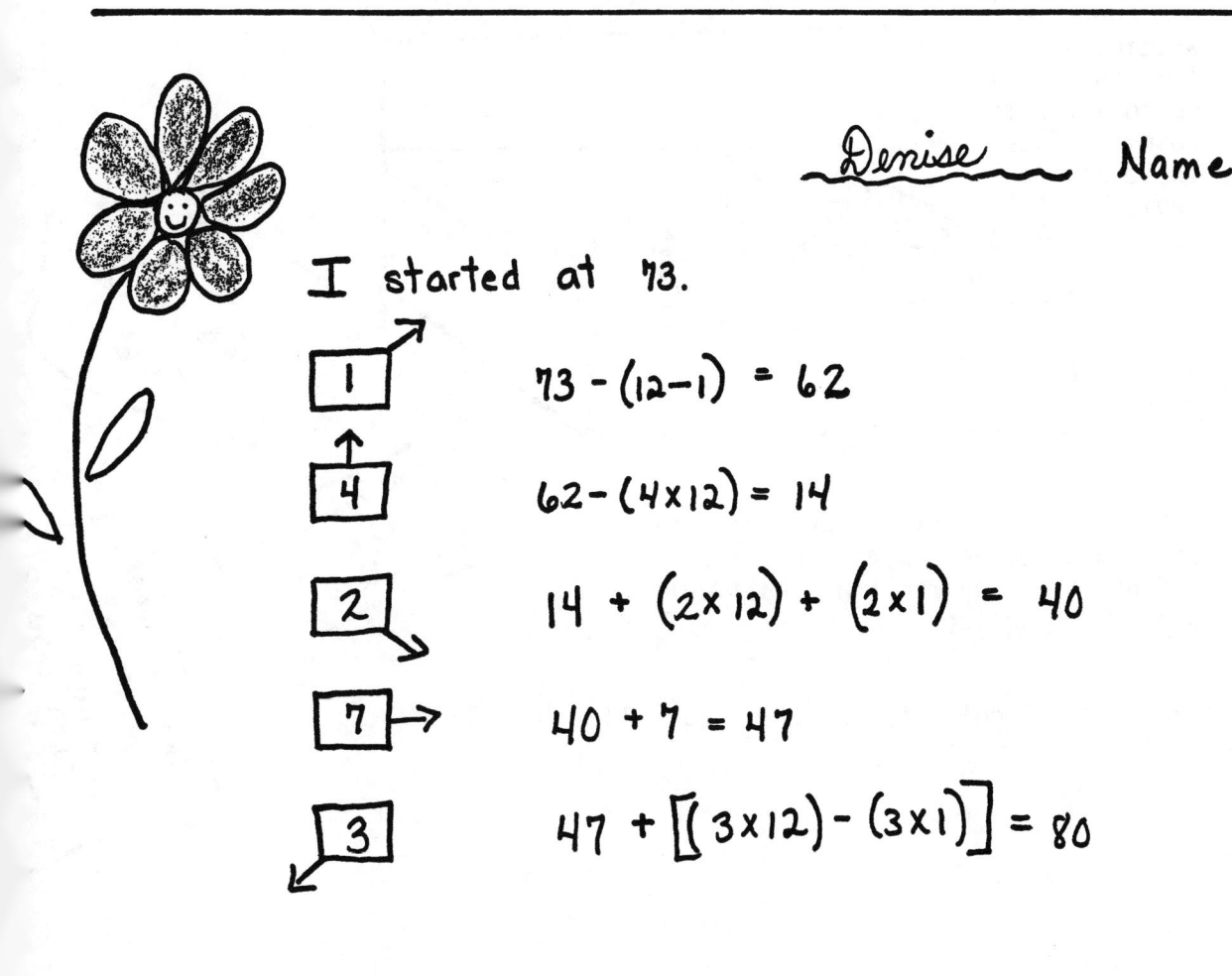

---

_Denise_  Name

I started at 73.

| 1 ↗ | $73 - (12 - 1) = 62$ |
| 4 ↑ | $62 - (4 \times 12) = 14$ |
| 2 ↘ | $14 + (2 \times 12) + (2 \times 1) = 40$ |
| 7 → | $40 + 7 = 47$ |
| 3 ↙ | $47 + [(3 \times 12) - (3 \times 1)] = 80$ |

142

## SAY IT WITH COINS

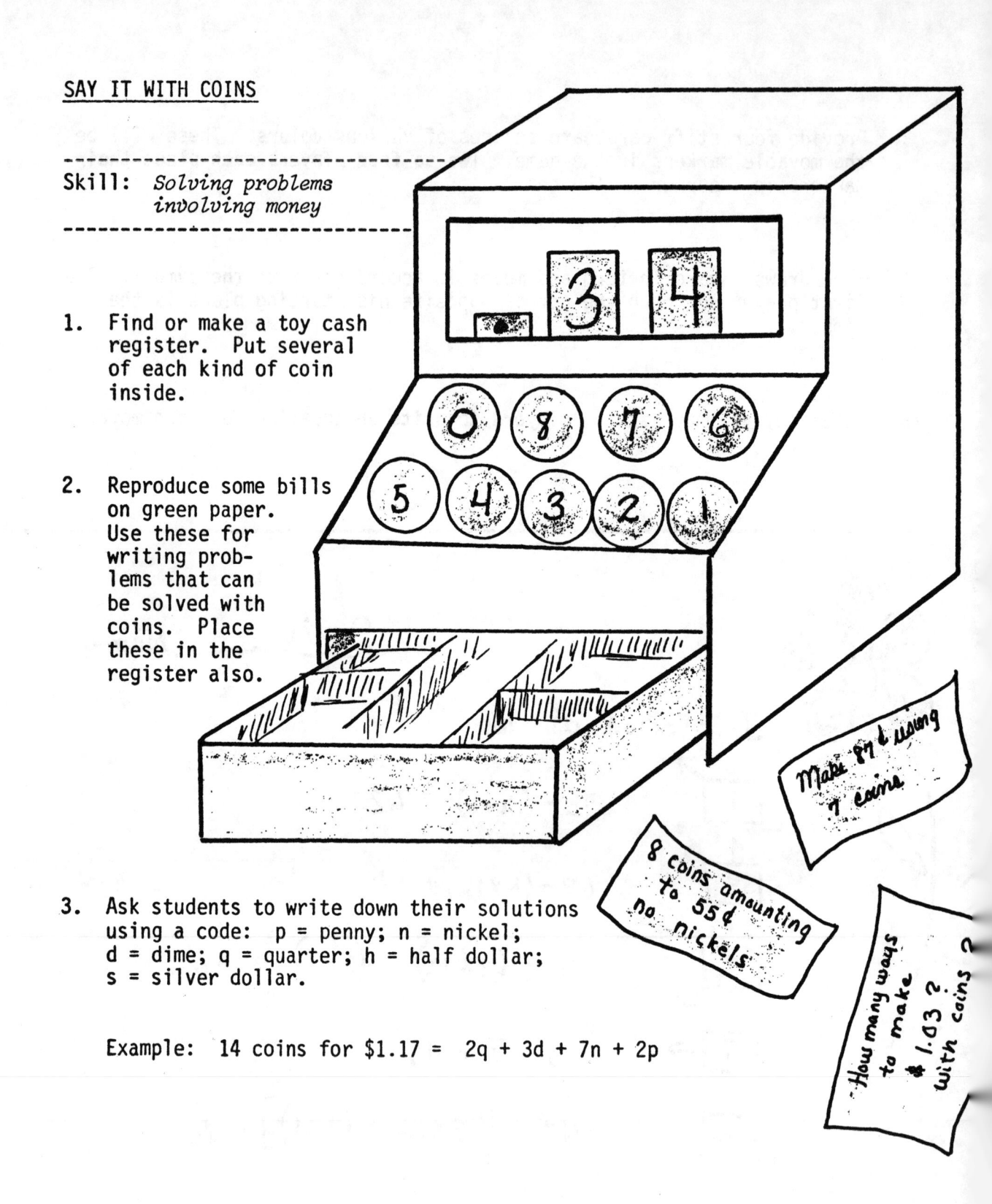

**Skill:** *Solving problems involving money*

1. Find or make a toy cash register. Put several of each kind of coin inside.

2. Reproduce some bills on green paper. Use these for writing problems that can be solved with coins. Place these in the register also.

3. Ask students to write down their solutions using a code: p = penny; n = nickel; d = dime; q = quarter; h = half dollar; s = silver dollar.

   Example: 14 coins for $1.17 = 2q + 3d + 7n + 2p

*Make 87¢ using 7 coins.*

*8 coins amounting to 55¢ no nickels*

*How many ways to make $1.03? with coins?*

SPORT SHOP

---
Skill: *Solving problems involving money*
---

1. Paint a Sporting Goods Shop on mural paper.

2. Ask students to add items to sell in the shop. Have them include price tags on their merchandise.

   *Buy 3 things for 25¢. What can you get?*

   *Buy 2 items totaling less than $10.00. What is the change?*

3. When the shop is ready for its grand opening, place numbered problems in an envelope near the entrance. These may be written on football, baseball, and basketball shapes. Include another envelope with blank shapes.

4. For each problem, the student writes his equation and solution on a blank football or baseball.

   *What will you pay for a new bowling ball and carrying case?*

   *What can you buy with $3.57?*

   *There is a 20% discount on all hockey equipment. How much will you pay for skates and a new stick?*

144

## NEED IT OR NOT?

---
Skill: *Determining what information is and is not essential to solving a problem*

---

1. Make up a story problem or take one from a book.

2. On narrow strips of paper, print each fact and question necessary for solving the problem. Add one strip with extra information which <u>won't</u> be useful in finding the solution.

3. Place all the strips in an envelope. Label the envelope "A."

4. Follow the same procedure with 15-20 more problems. In each envelope:

    . add unnecessary information,

    > OR

    . leave out one essential fact,

    > OR

    . leave out the question which the problem is to answer.

5. Place all the envelopes in an attractive holder directing students to examine the contents of each envelope and to decide what information is missing and what is extraneous.

6. Ask some students to write their own stories and add an envelope to the collection.

145

WHICH WITCH?

Skill: *Determining what operation is necessary to solve a problem*

1. On a bulletin board, place four witches with cauldrons. Make each cauldron from black paper backed by an envelope that will hold problems. Label each of the cauldrons either ADD, SUBTRACT, MULTIPLY, or DIVIDE.

2. On varying sizes of yellow and green bubbles, write story problems without numbers. Number each problem and put them into a fifth cauldron.

3. Students sort the problems according to operation and place them in the correct cauldron.

4. Provide an answer witch for self-checking.

SOMA THIS, SOMA THAT

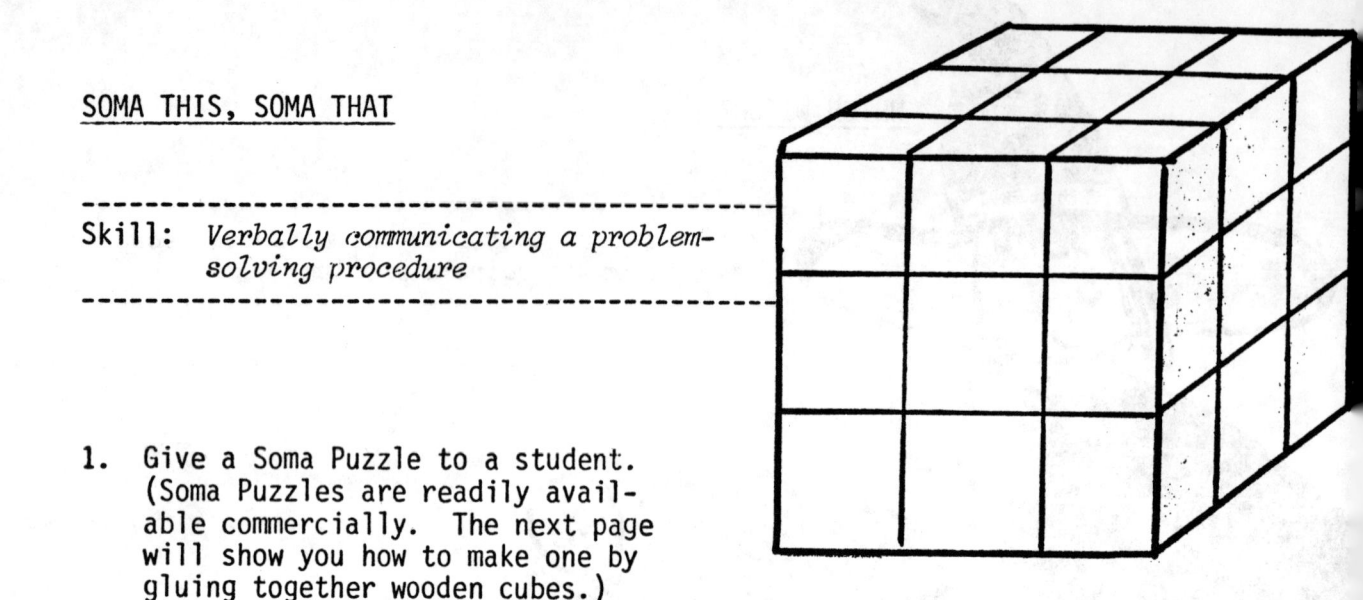

---
Skill: *Verbally communicating a problem-solving procedure*
---

1. Give a Soma Puzzle to a student. (Soma Puzzles are readily available commercially. The next page will show you how to make one by gluing together wooden cubes.)

2. After he has discovered and practiced a way to put the puzzle into a cube, ask him to instruct another student.

3. Place a screen or tall piece of cardboard between the two students. Make sure each one begins with the scattered pieces of like puzzles.

4. Student Number One explains step-by-step the procedure for putting the puzzle together.

Variation: This is a tough one for beginners! For easier tasks, try building other shapes with cubes or provide simpler puzzles or problems which one student can teach to another.

## HOW TO MAKE A SOMA PUZZLE

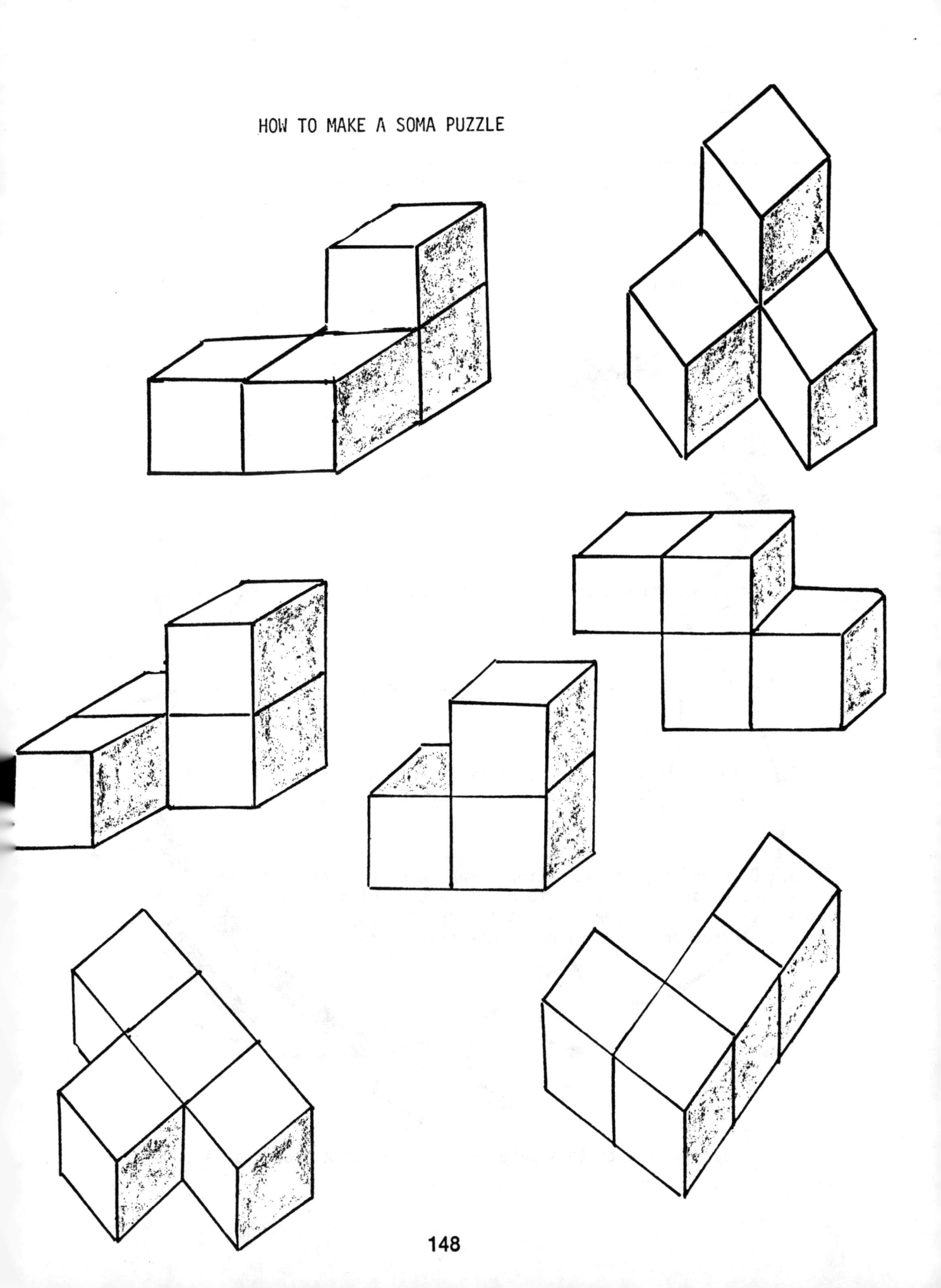

148

BUG-A-BOO

---
Skill: *Examining a problem solution for accuracy*
---

1. Have students paint and cut out large bugs (or anything else you might find in the sky--stars, birds, clouds, balloons, satellites).

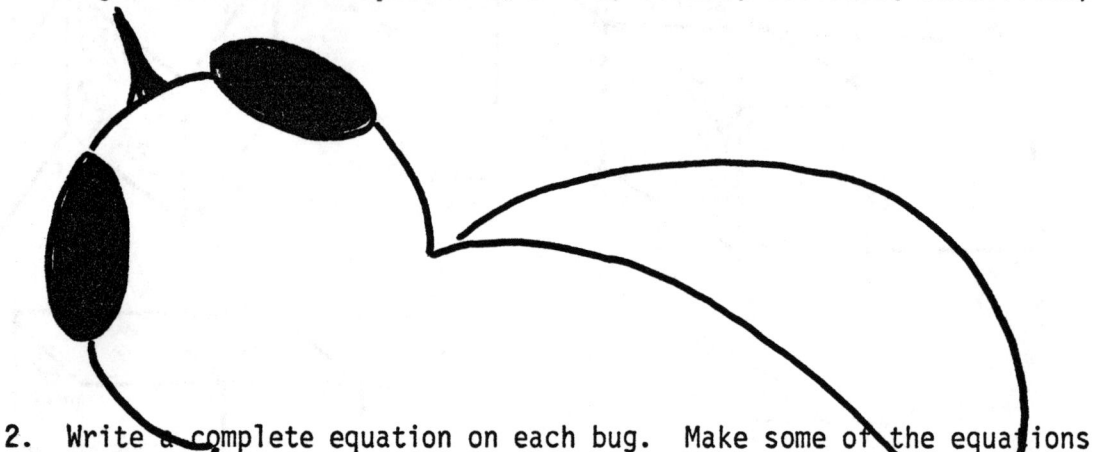

2. Write a complete equation on each bug. Make some of the equations incorrect.

3. Give a number to each bug and pin it flat on the ceiling.

4. Allow a period of two to three days in which students are to examine each problem in order to determine the accuracy of the equation.

5. On a sheet of construction paper, each student makes two columns--one in which to draw the bugs with correct problems, and one in which to draw the bugs with incorrect problems (with the mistakes corrected).

NOTE: The bug on this page may be enlarged and used as a pattern.

## ARE YOU QUICK ON THE DRAW?

**Skill:** *Recognizing that some problems have many solutions*

1. For this game players will need:

    --A watch with a second hand or a stopwatch

    --60 PLAY CARDS with digits 0-9

    --50 ANSWER CARDS with any numbers from 15-100

    --large pieces of construction paper

    --Instructions for playing the game

    --Someone to time the game

2. One or more students may play. Two may take turns or play simultaneously.

## HOW TO PLAY QUICK DRAW

1. *Draw an ANSWER CARD.*

2. *Begin drawing PLAY CARDS until you have cards that you can combine to get that answer.*

3. *You may use any operations you need.*

4. *Work on construction paper. You may want to write signs between the numbers.*

5. *Keep track of the time it takes to complete a play. That is your score.*

6. *Add one point for each extra card you drew.*

7. *Keep playing. Take a new ANSWER CARD.*

8. *After ten rounds, add your scores. The <u>lowest</u> score wins.*

## PASS THE BASKET

Skill: *Recognizing that some problems have many solutions*

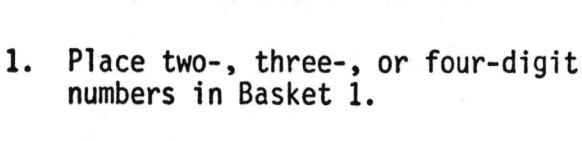

1. Place two-, three-, or four-digit numbers in Basket 1.

2. Place numbers 0-20 in Basket 2.

3. Provide paper for students to write equations.

4. Draw one number from Basket 1. Draw four numbers from Basket 2.

5. Tell students that they can use only those four numbers from Basket 2 to get the answer drawn from Basket 1. (They do not have to use all four, and can operate in any way with those numbers.)

6. After equations are written and shared, students may continue to draw numbers individually or in groups.

## WHAT'S MY RULE?

---
Skill: *Telling the rule for a function*

---

1. Find a sturdy box large enough to hold a person.

2. Have students paint or decorate it to look like a computer. Make an IN slot on one side and an OUT slot on the other side of the box.

3. Attach a pocket to the "machine." In the pocket, place 6 x 8 index cards which have been divided in half by a line.

4. Someone crawls inside the computer to act as its Brain. This person will decide on a function rule.

5. Students write numbers on the top halves of computer cards; they insert the cards into the IN slot; the machine applies the function and writes a second number on the bottom of each card; he returns the cards through the OUT slot.

6. When someone guesses the rule, he becomes the new "Brain" of the computer.


Variations:
1. Use shapes, letters, or other symbols. The "Brain" may reverse or change them according to a rule he creates.

2. Have students keep function tables showing the pairs of numbers for each rule. They can go on to anticipate how the machine will answer operating with a given function.

# TWENTY QUESTIONS

---
**Skill:** *Identifying unknown number by observing its properties in equations*
---

1. Tell the class that you are thinking of a number between 1 and 20. In order to find out what the number is, they may ask questions using it in problems.

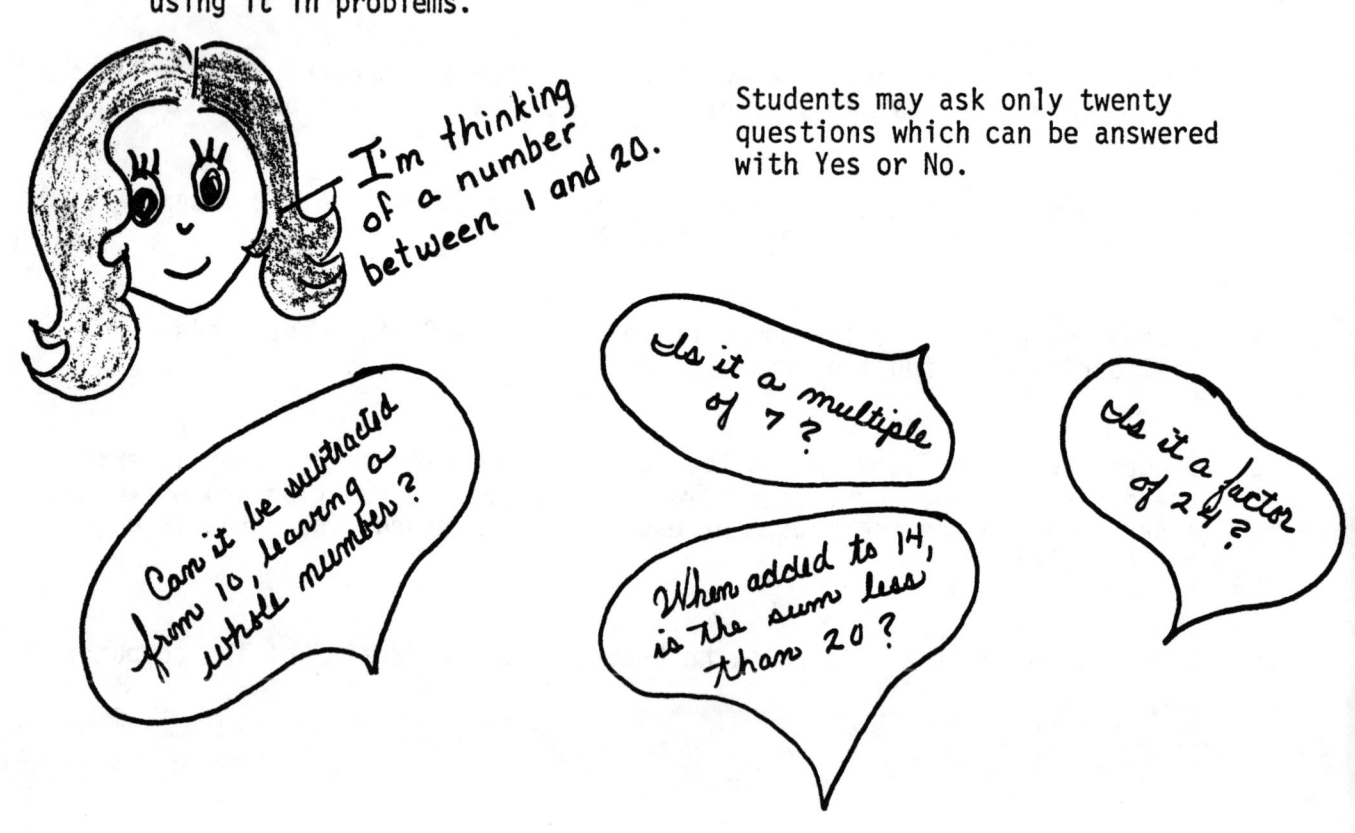

   *I'm thinking of a number between 1 and 20.*

   Students may ask only twenty questions which can be answered with Yes or No.

   *Can it be subtracted from 10, leaving a whole number?*

   *Is it a multiple of 7?*

   *When added to 14, is the sum less than 20?*

   *Is it a factor of 24?*

2. When someone guesses the number, let him choose another number.

3. Split the class into pairs or small groups and let them continue the game.

## BALANCE BOX

---
Skill: *Writing equations using letters to represent unknown quantities*
---

1. Find or make a small cup-type balance or scale.

2. Gather a number of objects small enough to be weighed in the balance:

   | paper clips | thumbtacks | paper fasteners | rings |
   | Q-tips | pins | erasers | chalk |
   | rubber bands | small blocks | staples | beads |
   | marshmallows | jelly beans | bottle caps | pennies |

3. Develop a code for the objects you have found:

   pc = paper clips
   t = tack
   p = pin
   ch = chalk, etc.

4. Write the code inside the lids of a shoebox. Store all the objects in the box as well. Include directions and problem sheets in the box.

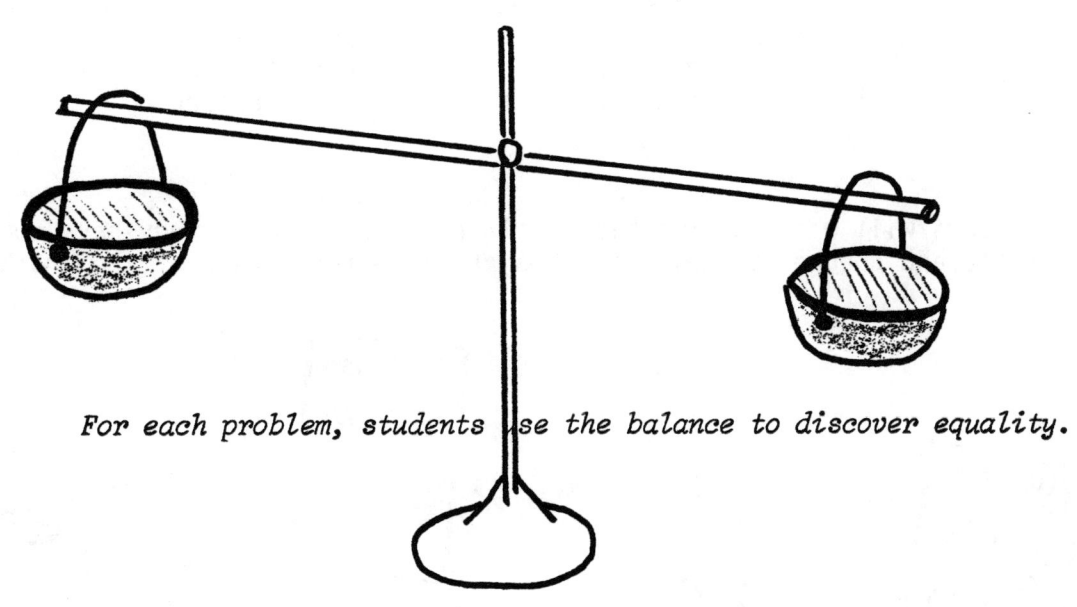

   *For each problem, students use the balance to discover equality.*

5. Students may be asked to complete equations such as:

   2Q = _____ p               m + j = _____ + _____

154

# TELL ME ABOUT THE BIRDS AND THE BEES!

**Skill:** *Solving problems which have two unknown quantities*

1. Make tables similar to those pictured below. At the top of each table, write an equation containing two variables. Cover these charts with plastic.

2. Cut several large birds and bees from sturdy posterboard. Glue one equation table to the front and back of each bird. (You may wish to color code the birds for difficulty of the equations.)

3. Punch holes in the birds. Suspend them from the ceiling on long pieces of yarn. Hang each bird on an S hook attached to the end of the yarn. This will allow students to remove a bird from its holder.

4. Students will write on the birds with a crayon. When they have completed and checked a table, they may replace the bird and take another.

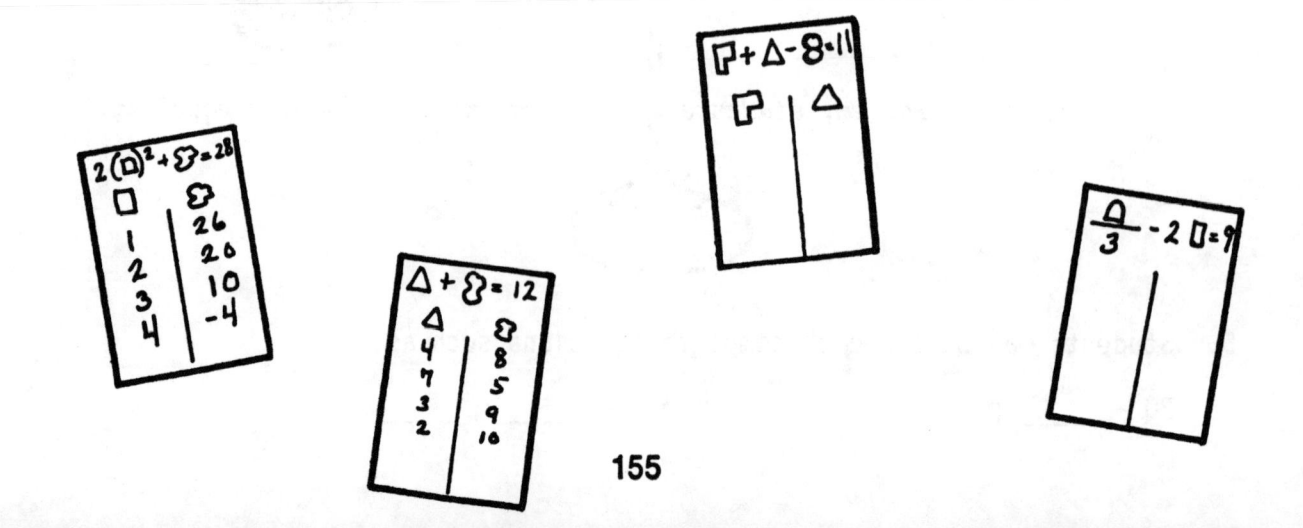

$2(\square)^2 + \vardiamond = 28$

| $\square$ | $\vardiamond$ |
|---|---|
| 1 | 26 |
| 2 | 20 |
| 3 | 10 |
| 4 | -4 |

$\triangle + \vardiamond = 12$

| $\triangle$ | $\vardiamond$ |
|---|---|
| 4 | 8 |
| 7 | 5 |
| 3 | 9 |
| 2 | 10 |

$\square + \triangle - 8 = 11$

| $\square$ | $\triangle$ |
|---|---|

$\dfrac{\square}{3} - 2\square = 9$

156

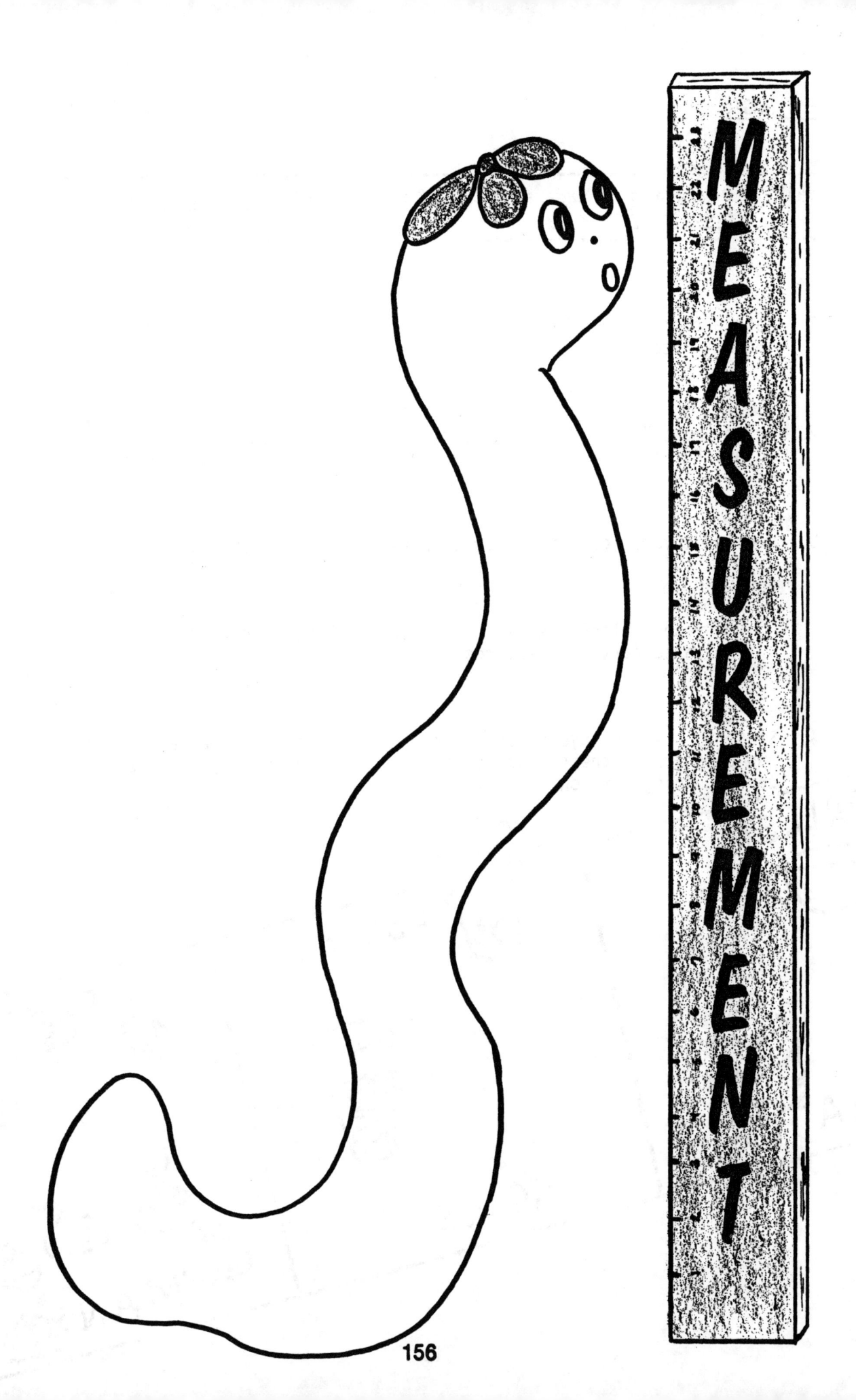

MAT MAGIC

---
Skill: *Demonstrating understanding of the concepts* more, less, smaller, greater, few, many
---

1. Make a concept mat from a piece of felt about one yard square. Divide the mat into three sections.

2. Write "more," "less" in one section; write "greater" and "smaller" in another section; write "many" and "few" in the third section.

3. Provide several objects or containers of objects for comparison.

   For example:

           beads, blocks, shoes, assorted bottles, coins, cans, waxed cartons, pictures, etc.

4. Students use the objects and containers to compare sizes and amounts and demonstrate their understanding by placing the objects beside the corresponding word on the mat.

# TIME IS ALIVE!

---
Skill: *Telling time*
---

1. Draw or tape a large circle on the floor (or use a circular rug).

2. Pin one of the numbers 1-12 on each of twelve students.

3. Ask them to place themselves on the circle to make a clock.

4. Choose two other students (one short, one tall) to be the hands of the clock.

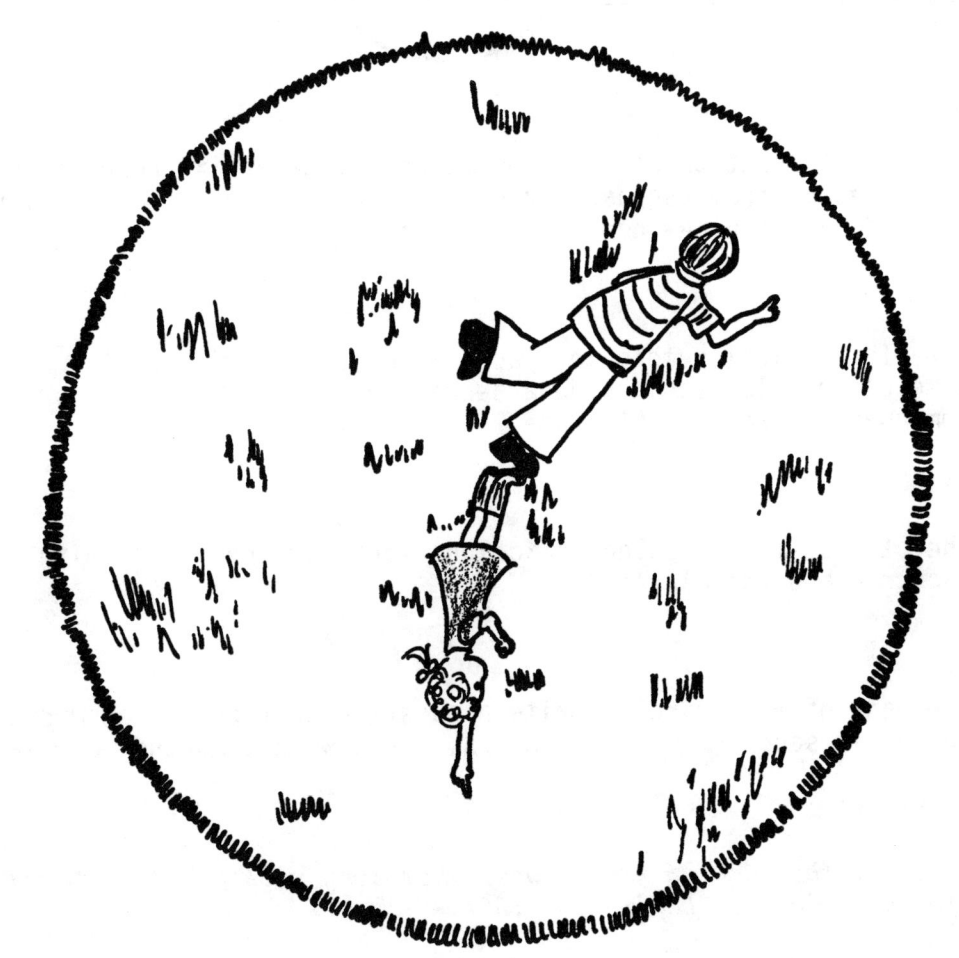

5. The students representing the "hands" lie on the floor with their feet in the center, positioning themselves to show the time as you dictate.

6. Allow other students to take turns being the "hands" as you continue dictating the time.

## HOW LONG DO I HAVE TO WAIT?

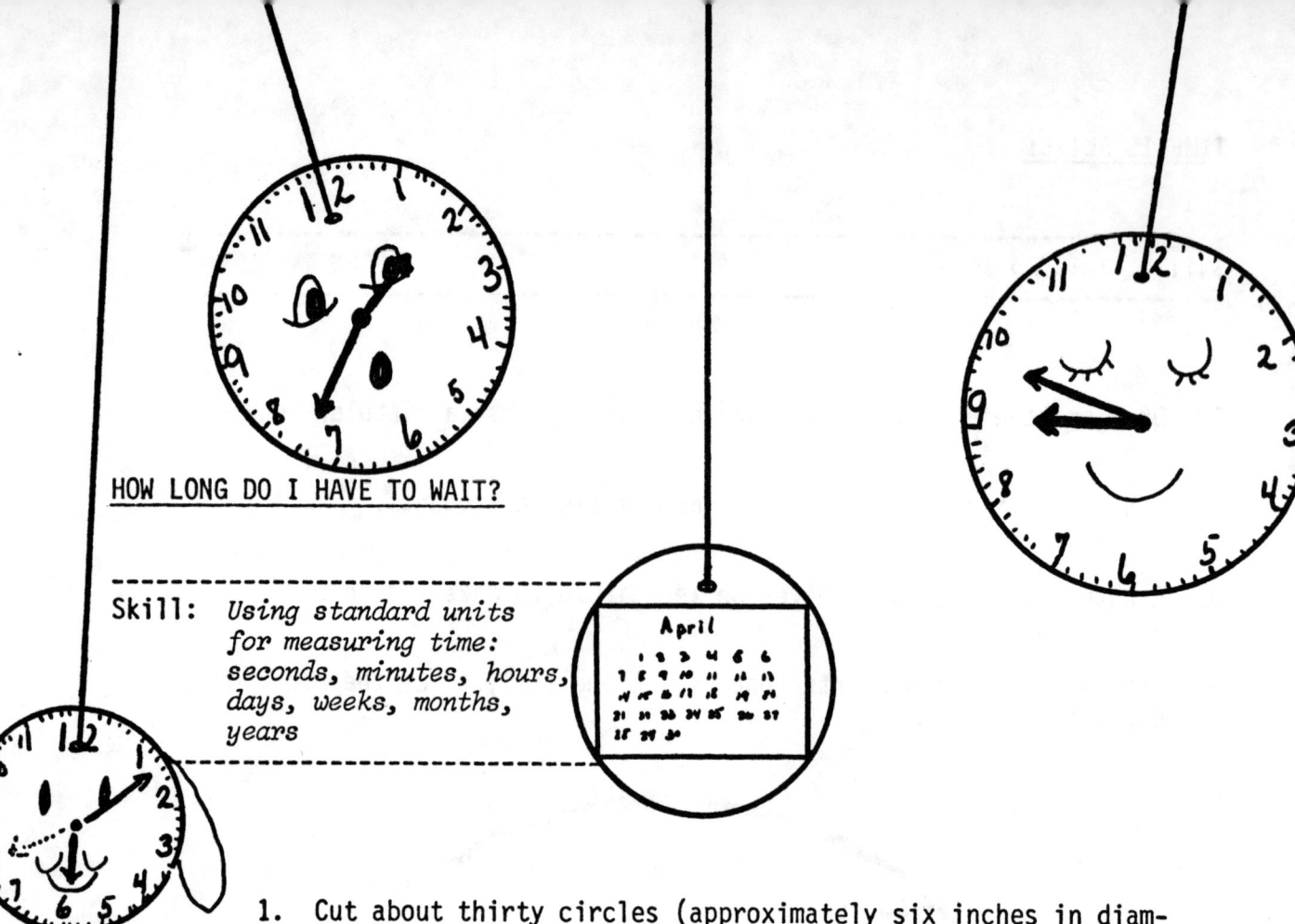

**Skill:** *Using standard units for measuring time: seconds, minutes, hours, days, weeks, months, years*

1. Cut about thirty circles (approximately six inches in diameter) from tagboard or sturdy construction paper. Or, you may use nine-inch paper plates.

2. Make half the circles into the faces of clocks showing different times. On some of the clocks, include a second hand. Specify a.m. or p.m. in some way on each clock's face.

3. On the others, make a calendar for one month or one year. With a red pen, circle a particular day.

4. On the back of each circle, write a question that can be answered by measuring in seconds, minutes, hours, days, weeks, months, or years.

   For example:

   a. *Look at the time on this clock. How long will you have to wait for a movie that begins at 7:30?*

   b. *Check the date on this calendar. How long will you have to wait for your paycheck on the first of next month?*

   c. *Look at the calendar! How long will you have to wait for Christmas, 1978?*

# "GRANDFATHER" KNOWS BEST!

Skill: *Measuring time using seconds, minutes, hours*

1. Paint a tall grandfather clock on a long piece of mural paper, or make a clock from cardboard.

2. Make movable hands on the face of the clock.

3. Ask a student to show 3:50 on the clock. Take turns using the clock to show various times.

4. Supply a list of activities and questions such as:

   a. Show 12:18. What time will it be in 57 minutes?

   b. It is now 4:13 p.m. What time will it be in 3½ hours?

   c. How long is it from 1:37 a.m. until 6:50 p.m.?

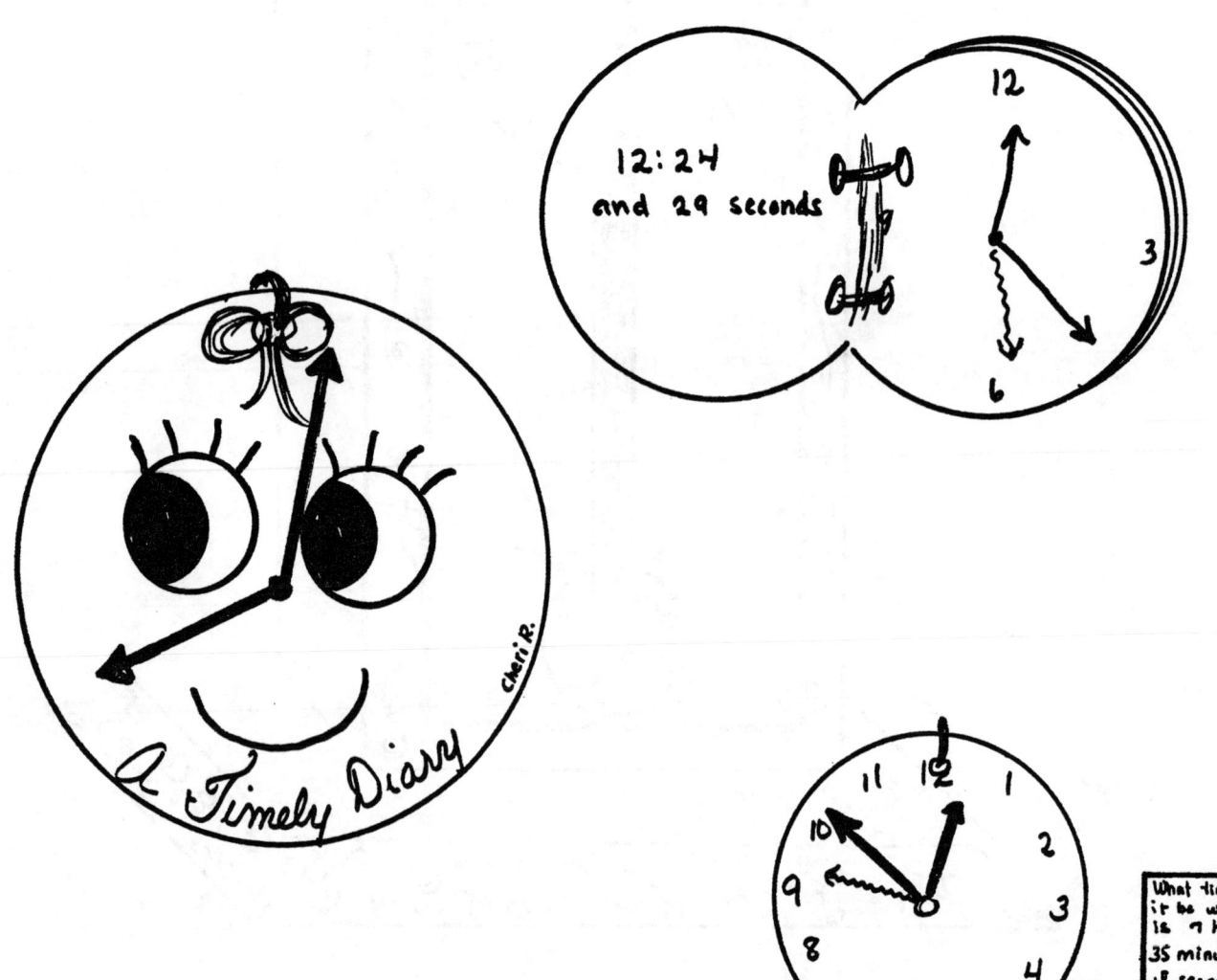

# SERVICE THE JET SET

---
Skill: *Solving time problems involving the earth's time zones*

---

1. Set up a world travel service in a corner of the room. Display a world map (labeled with time zones). Provide other maps, rulers, airline and ship schedules. Decorate the center with travel posters, airline ads, etc.

2. Make a list of approximate flying times between the main cities of the world.

3. Write several tasks which require students to prepare travel schedules between cities in different time zones.

4. For each trip planned, the student writes a ticket or series of tickets telling the estimated arrival time. These are some possible trips:

   a. *The Barker family lives in Los Angeles. They wish to leave February 17 in the afternoon for Cairo. Plan their trip. What time will they arrive?*

   b. *Mr. Chung is a businessman from Hong Kong. He wishes to leave home next Monday morning and spend four to five hours in each of these cities: Madrid, Tokyo, San Francisco, Honolulu, Chicago, Munich. He must return home by Saturday. Plan his trip. Tell when he will arrive at each destination.*

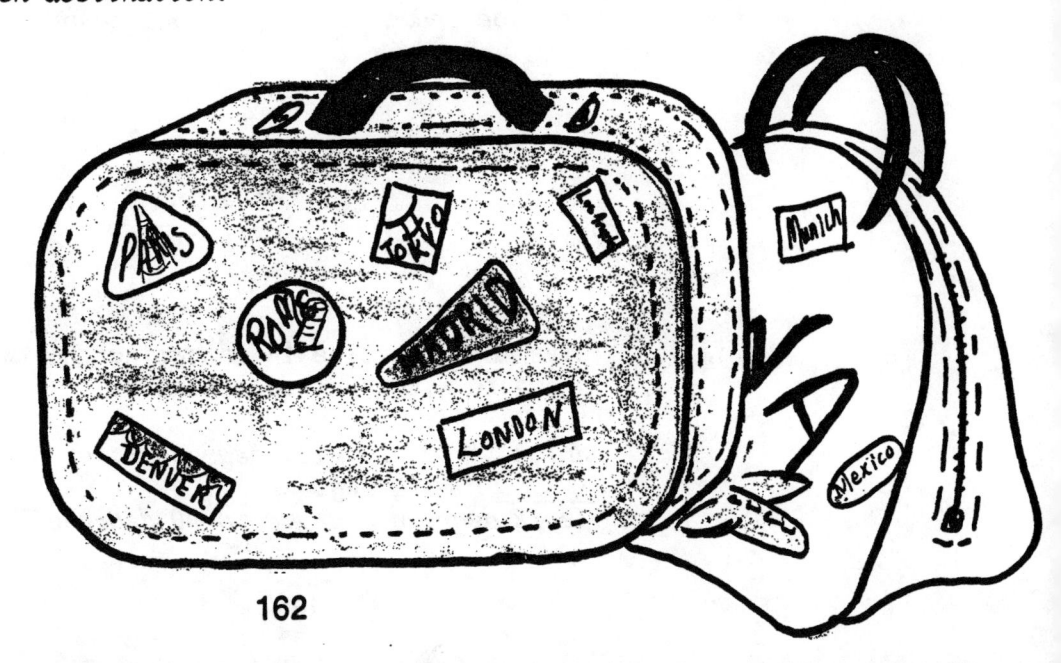

## HOT IS HOT

---
**Skill:** *Using a Celsius and Farenheit thermometer to measure temperature*

---

1. Compare a Celsius and a Farenheit thermometer. Point out the difference in size of the degrees, freezing point, and boiling point.

2. Ask students to measure some temperatures using both thermometers and compare the results.

   °Farenheit = (°Celsius x 1.8) + 32

   °Celsius  = (°Farenheit - 32) x .556

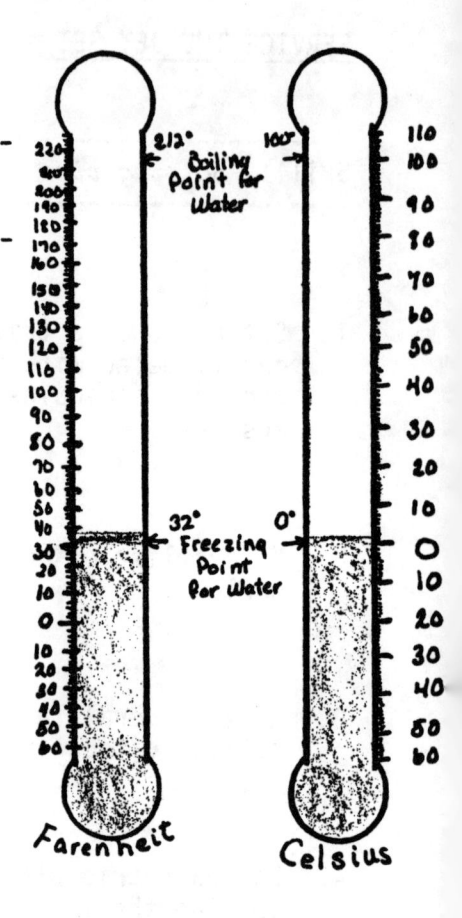

### Ideas!!

| | |
|---|---|
| room temperature | drinking fountain water |
| milkshake | temperature outdoors |
| boiling water | temperature in kitchen |
| my ice cream | Add your own: _____ |

**Variation:** Give more practice in comparing the two units of measurements with such questions as:

*Which is lower: 110 degrees F. or 38 degrees C?*
*____? F = 72 degrees C.*
*Is 72 degrees C close to 90 degrees F?*

WORM YOUR WAY

Skill: *Finding lengths using centimeters, meters, inches, feet, and yards*

1. From tagboard, make a "measure worm" for each unit: centimeter, meter, inch, foot, yard.

2. Make a list of distances to be measured. Duplicate them on a Wormy Work Spot. You may wish to reproduce the following page.

3. Ask the students to find each length by using one or more the five units. They may round to the nearest half-unit.

    For example:

    *My pigtail is 1 foot, 3½ inches long.*

    *Julie is 1 meter, 52½ centimeters tall.*

    *My ear is 6½ centimeters long.*

    *It is 5 yards, 2 feet and 7½ inches to the drinking fountain.*

    *Mr. Black's left leg is 1½ yards long.*

# YOUR WORMY WORK SPOT

- length of your ear
- width of the blackboard
- distance to the gym from our door
- width of the flag
- (use one of the unit worms for each measurement. You add one!) →
- height of your best friend
- length of the cafeteria
- height of a swing (from the ground)
- height of your knee (from the floor)
- distance around the classroom
- height of the classroom
- length of the longest leg in the school
- distance from your desk to the door
- find and record each of these measurements.
- height of the drinking fountain
- the principal's height
- length of your thumbnail

# MYSTERY MESSAGE

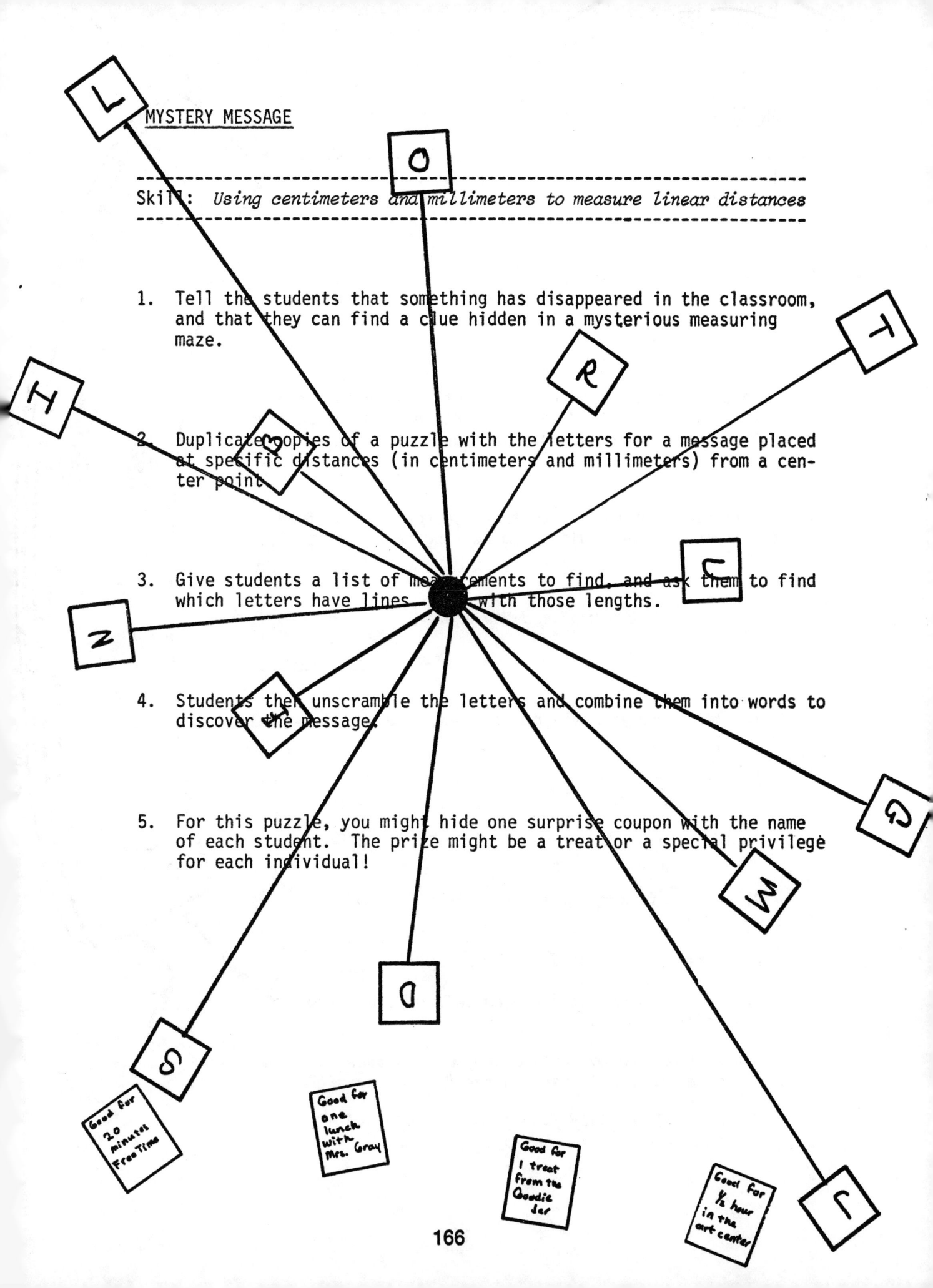

---
Skill: *Using centimeters and millimeters to measure linear distances*
---

1. Tell the students that something has disappeared in the classroom, and that they can find a clue hidden in a mysterious measuring maze.

2. Duplicate copies of a puzzle with the letters for a message placed at specific distances (in centimeters and millimeters) from a center point.

3. Give students a list of measurements to find, and ask them to find which letters have lines with those lengths.

4. Students then unscramble the letters and combine them into words to discover the message.

5. For this puzzle, you might hide one surprise coupon with the name of each student. The prize might be a treat or a special privilege for each individual!

Good for 20 minutes Free Time

Good for one lunch with Mrs. Gray

Good for 1 treat from the Goodie Jar

Good for ½ hour in the art center

SCAVENGER HUNT

---
Skill: *Using standard linear measurements*
---

1. Make up a scavenger hunt which will require students to find things in the room by measuring linear distance accurately.

2. Give the students a direction sheet similar to the one below. They must find each measurement and record the name of each object as they locate it.

3. You might offer a prize to those scavengers who find all the items (a new ruler or measuring tape, perhaps?).

*HOW MANY CAN YOU FIND?*

*Name_____*

1. *Two yards from the pencil sharpener you will find a blue _____.*

2. *There are five _____ lying 2.5 centimeters from the upper right-hand corner of the teacher's desk.*

3. *What is located approximately 3 inches below your nose?*

4. *_____ is hanging 1 meter from the tip of the flag.*

5. *Three meters inside the doorway is ____.*

6. *Seven and one-half inches left of the principal's mouth is _____.*

7. *On your dictionary's cover, what letter is 17 millimeters directly beneath the "B" in the title?*

8. *One and one-half feet from the left end of the art supply shelf is a box of _____.*

9.  There is a treat for you 7 yards from the reading rug. What is it? _____

10. Whose classroom is 30 meters to the right of ours?

11. Which of your textbooks is 27 centimeters long?

12. What do you find 6 yards, 2 feet, and 7 inches down the wall from Top Cat? _____

Measure and compare the volumes of objects by discovering how much water each will displace. Place the object carefully in a container full of water; then catch and measure the water that overflows.

Collect scraps of carpet from students and "scrounge" carpet samples from retailers. Students will get plenty of practice measuring perimeters and areas as they plan and make a rug for a section of the classroom. Carpet tape secures the pieces together quite nicely.

Create a permanent measurement corner in the classroom. Collect as many measuring tools as you can.

Suggestions:

| rulers | measuring spoons | calendars |
| yardsticks | measuring cups | clocks |
| metersticks | pint containers | thermometers |
| tape measures | quart containers | assorted scales |
| protractors | gallon containers | grid paper |
| square units | peck baskets | watches |
| cubic units | bushel baskets | calipers |
| compasses | liter containers | maps |

Hang two measuring tapes--one English, one metric. Have students mark their heights and check their growth every month.

168

## PERIMETER PROBE

**Skill:** *Finding perimeters of closed figures*

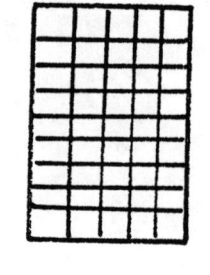

1. Divide large sheets of colored tagboard into two-inch square segments.

2. From these sheets, cut twenty-thirty groups of squares. Make them of varying shapes and perimeters. Cover them with plastic for durability.

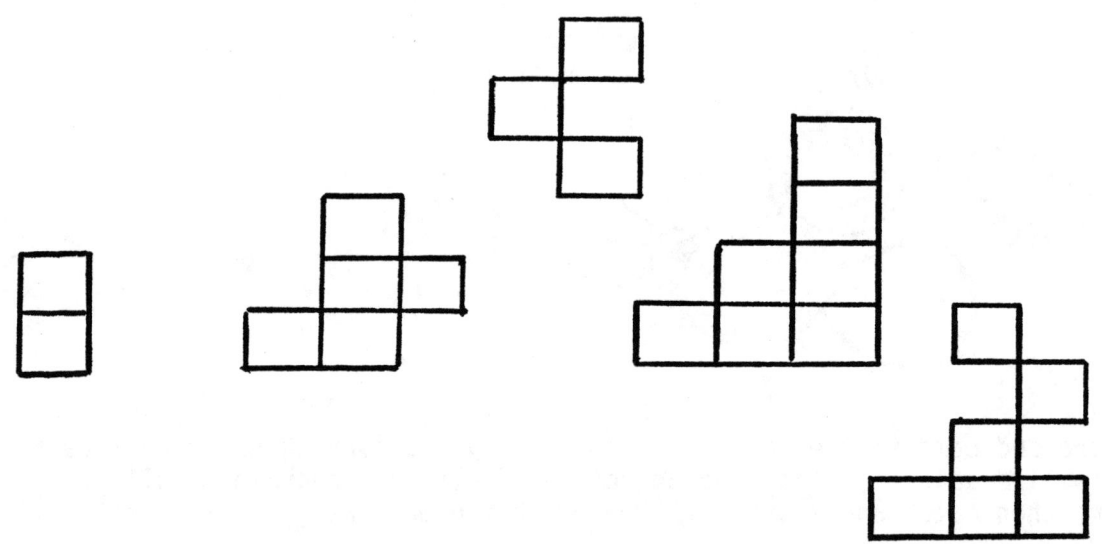

3. Provide a flat surface for easy arranging of the shapes (a gameboard, table top, piece of plexiglass, etc.).

4. Use individual squares for writing directions. You might color code the Instruction Squares according to difficulty.

5. Supply students with large pieces of paper for drawing pictures of some of their answers.

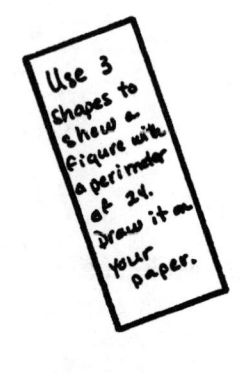

---

Use 3 shapes to show a figure with a perimeter of 24. Draw it on your paper.

Classify all the shapes by perimeter. Put all those with a perimeter of 8 in one pile, etc. ☺

Put all the shapes in order according to perimeter size.

Find the perimeter of each red shape.

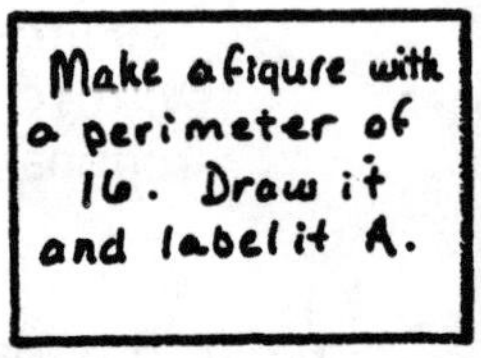

Make a figure with a perimeter of 16. Draw it and label it A.

# CURVE COLLAGE

---
Skill: *Finding the perimeter of closed figures*
---

1. Draw a collage of closed curves, making each figure with lines of a different color.

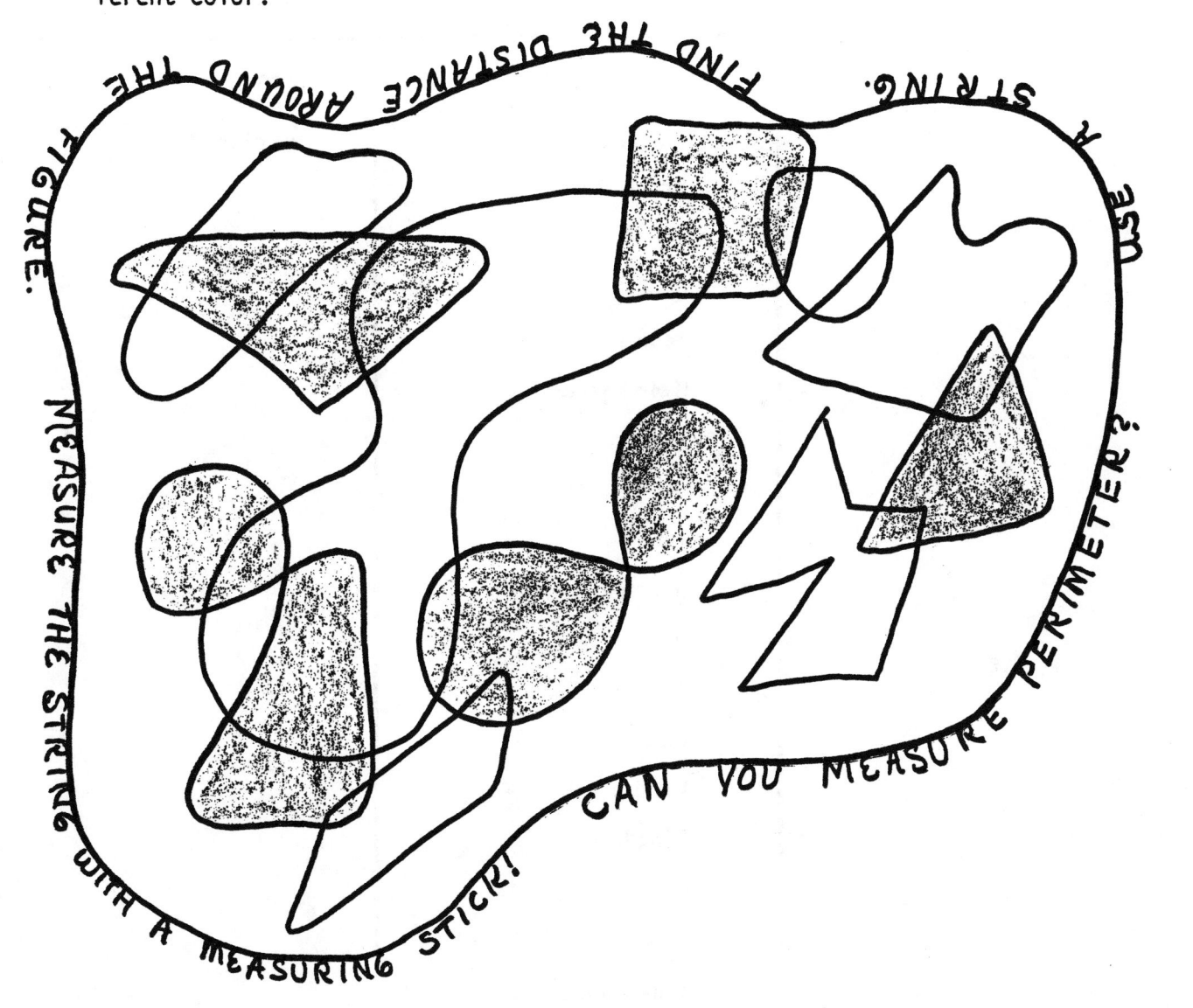

*FIND THE DISTANCE AROUND THE FIGURE. USE A STRING. CAN YOU MEASURE PERIMETER? MEASURE THE STRING WITH A MEASURING STICK!*

2. Give each student a measuring stick and a length of string or yarn (at least as long as the longest perimeter in the collage).

3. Direct him to find the perimeter of each shape by laying his string along the lines that make the figures. He then uses the measuring stick to determine how much string was used for each.

170

4. On his record sheet, the student notes the color and perimeter for each figure.

```
                              Name _____
         Color                 Perimeter
    light   blue
    dark   blue
    aqua
    green
     lime green
    purple
    lavendar
    red
    orange
    yellow
    silver
    gold
    brown
    black
    pink
    grey
    red-orange
```

Variation: Let students make original collages by turning the activity around. Give the color and perimeter--and each draws a figure of his choice to fit the description. Even given the same directions, the students will all produce different designs!

171

# WAS PYTHAGORAS "SHORT"-SIGHTED?

---
Skill: *Using the Pythagorean Theorem to find the length of one side of a right triangle*

---

1. Use chalk or tape to mark several right triangles on your classroom floor. Make the sides of the triangles lengths for which students can easily find the squares: 3 ft., 4 ft., 5 ft., etc. Label each triangle with a number.

2. Help students discover the Pythagorean Theorem-- $a^2 + b^2 = c^2$ in a right triangle with c being the hypotenuse, or side opposite the right angle. See illustration.

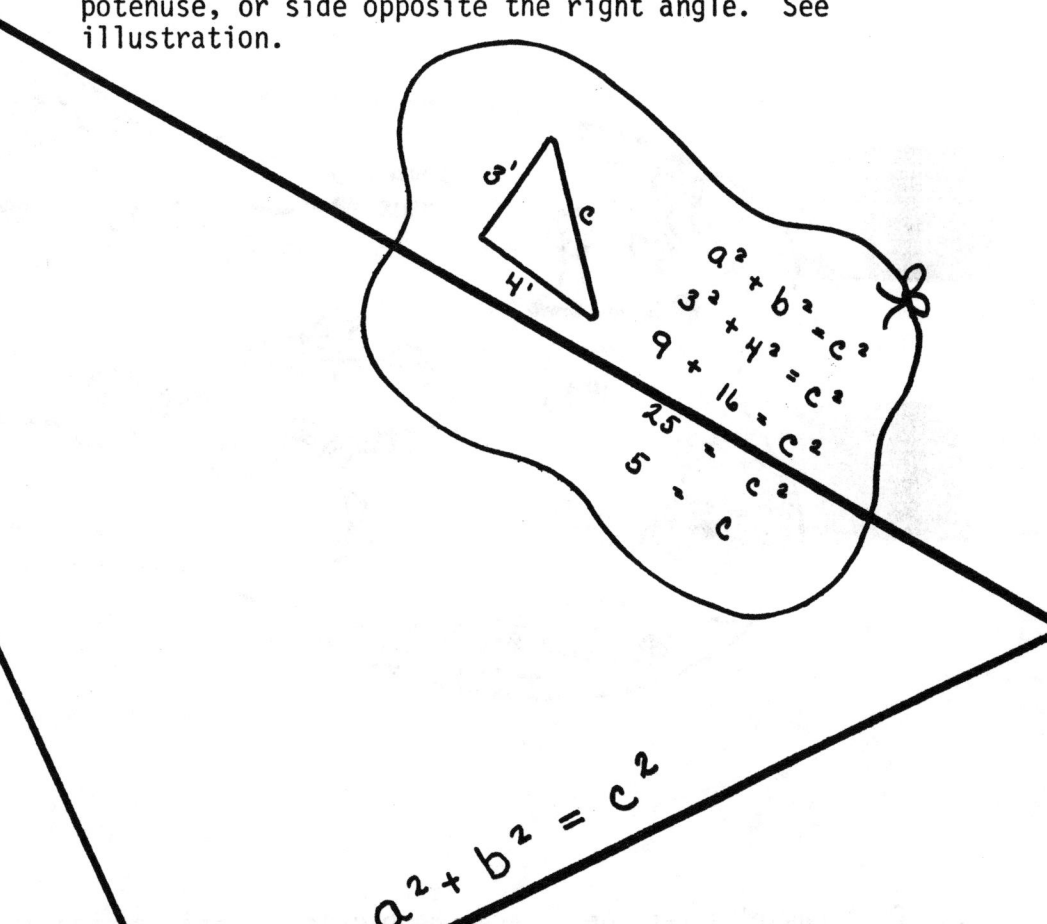

3. Have students find the hypotenuse of each triangle by measuring the two shorter sides and using the formula. They should check their answers by measuring the hypotenuse.

4. Students might use the sides of the triangles as paths for traveling about the room. Before visiting a friend, a student should figure out which route will be the shortest. Students will soon discover that the hypotenuse route is always shorter than walking the other two legs of the triangle.

172

# TREASURE MEASURE

---
Skill: *Using a map scale to measure distances*
---

1. Make a treasure map on a large piece of tagboard. Draw a red dot near each land form or point of interest on the map. Give the map this scale: 1 centimeter = 1 mile. Cover the map with plastic.

2. Place a centimeter ruler or meter stick, crayons, and a cassette player near the map.

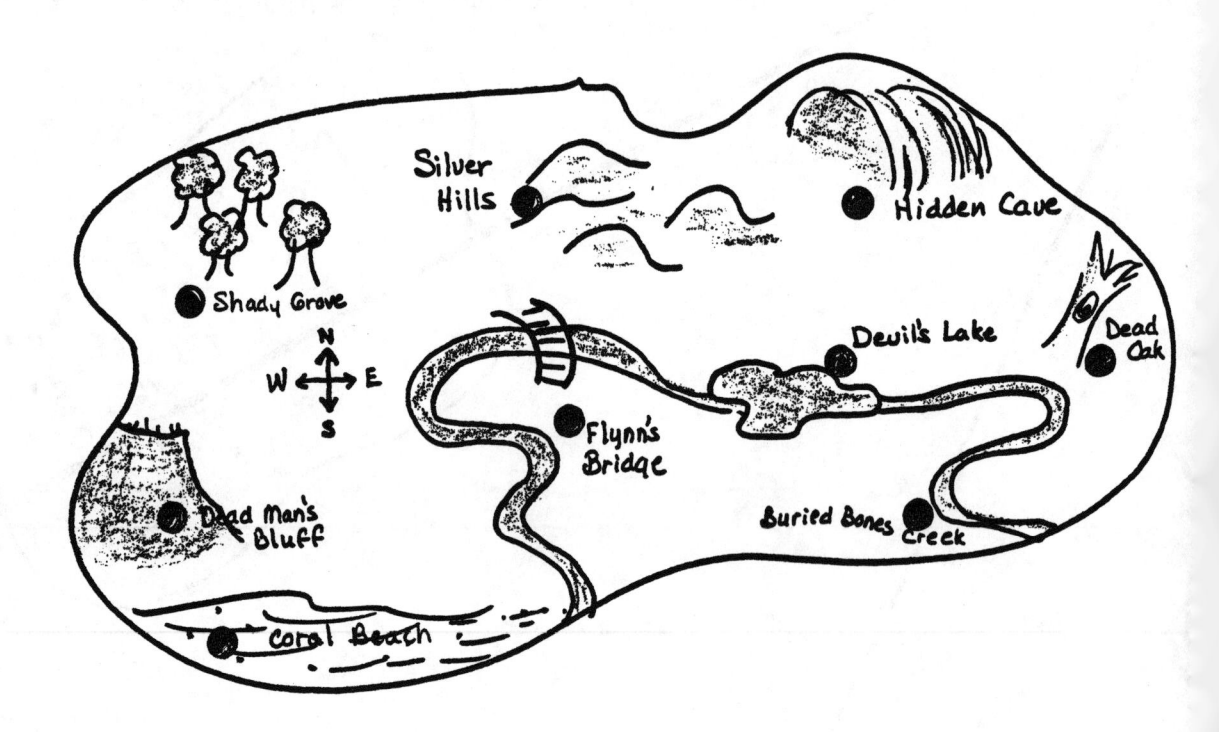

3. On a cassette recorder, tape the background information and directions for locating the treasure.

   The narration for your map should sound something like this:

   *"You are the only one who has reached Treasure Island. Can you find the treasure?"*

173

*Your landing is a spot eight miles from the southwest corner of this map. Can you find that spot? (Always measure to a red dot. Draw a line to that red dot.) Where have you landed?*

*Walk 32 miles northeast. Rest there for the night. Where have you stopped?*

*Your second day takes you 24 miles due east. Record your second stop!*

Continue the narration in this manner until one measurement reaches a spot you have chosen for the treasure.

4. The student measures from one point to the next, marking his route on the map with a crayon, stopping the recorder when he needs thinking or measuring time. A worksheet such as this one will help him record his search:

Keep A Record Of Your Treasure Hunt Here:

| | | Distance You Measured | Where you landed |
|---|---|---|---|
| START | SW corner of map | | |
| STOP 1 | | | |
| STOP 2 | | | |
| STOP 3 | | | |
| STOP 4 | | | |
| STOP 5 | | | |
| STOP 6 | | | |
| STOP 7 | | | |
| STOP 8 | | | |
| STOP 9 | | | |

WHERE AM I?

---
Skill: *Using lines of latitude and longitude to locate positions on the earth's surface*

---

1. Hang a world map on an easel or chart holder.

2. Cover the map with plastic so that students may write on it with crayon.

3. In the tray of the easel, place four cans or other containers labeled

   a. WHERE AM I?
   b. WHERE MIGHT I BE?
   c. HERE I AM!
   d. COULD I BE?

4. On small slips of paper, write positions to be located as follows: (Give each question a letter code.)

   For WHERE AM I? write a location in degrees, latitude and longitude.

   > A WHERE AM I?
   >
   > A city at
   >   0° latitude
   >   78° W longitude

   For WHERE MIGHT I BE? give only latitude or longitude (allowing students the possibility of choosing a point anywhere along that line).

   > G WHERE MIGHT I BE?
   >
   >   a country
   >     90° E longitude

For HERE I AM! give the name of a city or country (letting the student read the location).

> R
> HERE I AM!
> Warsaw, Poland
> Give the location.

For COULD I BE? give a latitude or longitude and the name of a place (leaving the student to decide if one could be in that city if on that line of latitude or longitude).

> V
> COULD I BE?
> 30° S latitude
> Could I be in Brazil?

4. The student uses a crayon to mark the positions he has located. He might also record his answers in a notebook.

ANGLERS' PARADISE

Skill: *Using a protractor to measure angles in degrees*

1. Supply plastic protractors for students or reproduce one that can be mounted on tagboard and cut out by each student.

2. Make a large picture mural that includes many angles. Label the angles with letters.

   TRY: a camping scene with tents, hats, firewood, axes, etc. forming different angles

   OR

   a town with many angular roofs

   OR

   a bake shop where pies and cakes are cut into pieces

3. Near the mural, hang an instruction sheet naming angles to be measured and questions to be answered.

4. Students may keep their answers on a separate record sheet.

   Variation: This activity can be expanded to include identification of kinds of angles.

177

# WEIGHT WATCHERS

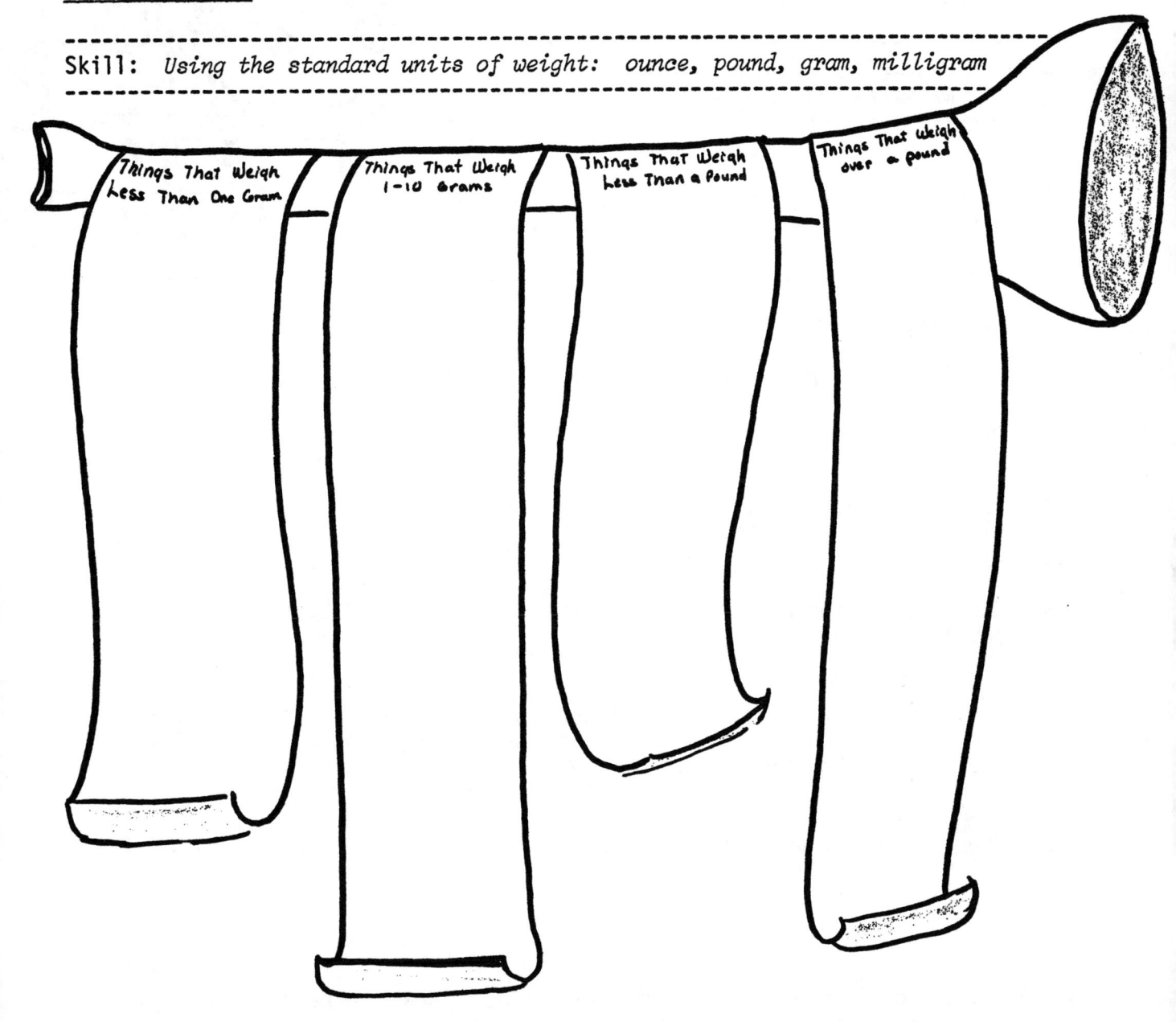

**Skill:** *Using the standard units of weight: ounce, pound, gram, milligram*

Things That Weigh Less Than One Gram

Things That Weigh 1-10 Grams

Things That Weigh Less Than a Pound

Things That Weigh over a pound

1. In your measuring center, hang four long pieces of burlap or mural paper with the headings shown above.

2. Ask students to write or pin onto the burlap names of objects which can be classified in one of the groups. The object itself or a picture may be hung as well.

3. The exact weight may be included with each object.

## LET'S PLAY POST OFFICE!

**Skill:** *Measuring, using standard units of weight: grams, ounces, pounds*

1. Transform a corner (or a large appliance box) into a post office. Gather supplies such as:

   scales (to measure in metric and English systems)
   markers
   wrapping paper
   zip code books
   postal rate charts (national and international)
   rubber stamps
   maps

2. Set up a series of mailboxes (one for each student). Try to hook other classes on the idea too!

3. Students may represent different cities in the United States or other countries.

4. Send letters, finished papers, treats, packages, gym shoes, etc. back and forth between the towns and countries. Everything that is exchanged must be taken to the post office for weighing and stamping.

5. The class might print its own series of stamps; other classes could create different stamps; and individuals might design stamps for their particular countries.

   NOTE: You'll find a lot of math skills being used in reading tables, figuring rates, and "selling" stamps!

SEARCH THE AREA!

---
Skill: *Finding the area of a closed figure*
---

1. Have students paint shapes on a long strip of paper (about 12 inches wide). Ask them to paint regular and irregular shapes.

2. Letter or number each region.

3. Hang the paper along a wall low enough so that students may sit on the floor beside it.

4. On plastic sheets (or blank overhead transparencies), draw grids of two sizes: centimeter squares and inch squares.

5. By placing a transparent grid over a region, the student can find the area in square inches or centimeters (or both). He keeps his own record of the measurements.

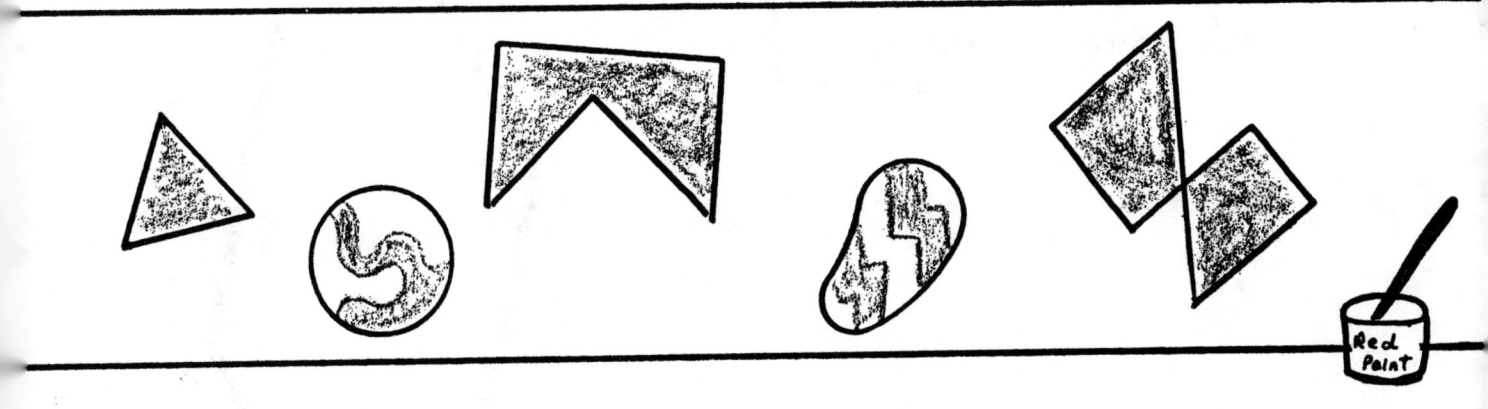

Variations:
1. Do a second "mural" and ask students to find the areas using only rulers (that is, without the help of the squares).

2. Give graph paper to students. Have them color regions (regular and irregular) for given areas.

180

## HOW DO YOU FIGURE THIS FIGURE?

---
Skill: *Finding the area of closed figures*
---

1. Have students make regular closed figures of yarn or string soaked in white glue. The shapes can be arranged on white construction paper. They will become stiff as they dry.

2. Watercolor paints may be used to paint the figures. Ask students to use color to distinguish between the interior and the exterior.

3. Display the figure in a center where measuring tools have been provided.

4. Display some figures you have made, in which you have written the formulas for finding areas.

5. Students find the interior area of each closed figure. Some of them might try to find the exterior area as well!

   (See Appendix for area formulas.)

   Variation: If students make irregular figures, the activity can become one in estimation of surface area.

BORED?  TRY GEOBOARDS!

---
Skill: *Finding the area of closed figures*

---

1. Supply six or more geoboards. If your school does not have any, you can make a geoboard easily by hammering nails at regular intervals (one inch) on a square board. Add rubber bands and you're ready to go! (Kids will enjoy pounding the nails and painting the finished products.)

2. Make a series of cards to accompany the geoboards. On each card, write a task which will require the student to find the area of a figure.

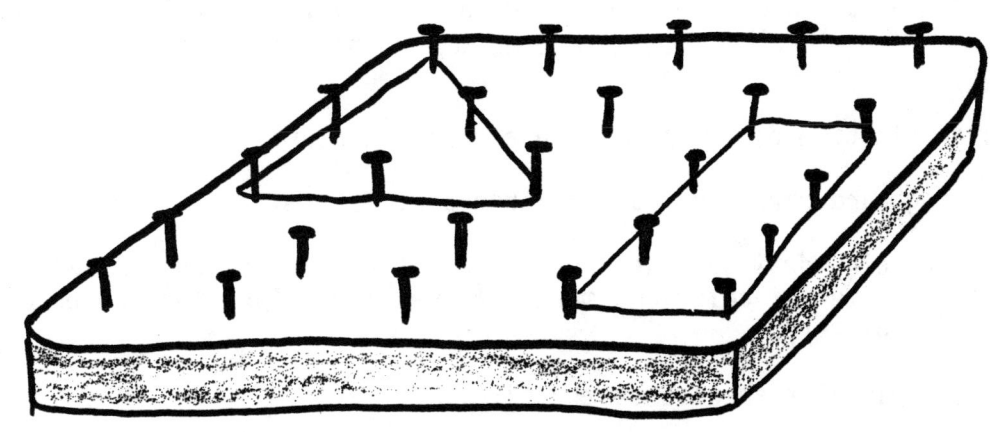

    Examples: *Make this figure on your geoboard. What is the area?*

    *Make a square with an area of 8.*

    *Make a rectangle that has an area of 3.*

    *Show a triangle with an area of 6.*

    *Show this figure. What is its area?*

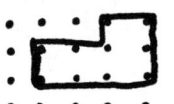

3. Provide dot paper so that the student may draw pictures of the figures he has made.

*Circumference = πd*

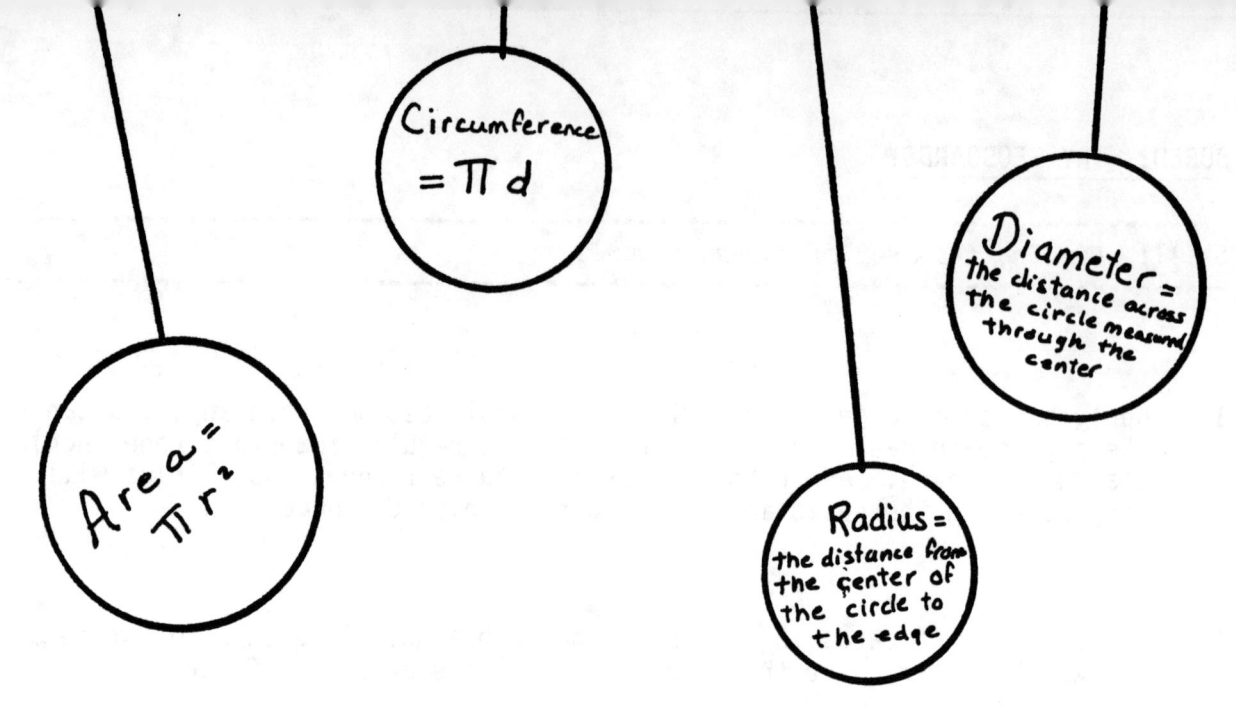

*Area = πr²*

*Diameter = the distance across the circle measured through the center*

*Radius = the distance from the center of the circle to the edge*

## CIRCLE CIRCUS

---
Skill: *Finding the diameter, radius, circumference, and area of a circle*

---

1. Find an old tire or two to use in a circle center.

2. Cut a tagboard circle almost as large as the tire.

3. On the circle, draw overlapping circles of various sizes. Draw each circle a different color and mark (in the same color) the center and a radius or diameter.

4. Tape the circular sheet onto the tire.

5. Hang these directions in the center of the tire on the opposite side. (See illustration below.)

*DIRECTIONS*
*Level I: Measure the radius of each circle.*
*Level II: Tell the diameter of each circle.*
*Level III: Find the circumference of each circle.*
*Level IV: Find the area of each circle.*

# FIGURE FOIL

**Skill:** *Finding the surface area of space figures*

1. Provide a collection of objects which are cubes, spheres, cylinders, cones, pyramids, or prisms.

2. Supply wrapping paper or aluminum foil, tape, and single-edge razor blades.

3. Students wrap figures, completely taping all loose edges of paper. (Aluminum foil will cling well without taping.)

4. When figures are well wrapped, instruct students to slit the paper carefully along enough edges to remove the wrapping in one flat piece.

5. Students then find the area of the figure by measuring the sections of the flat wrapping.

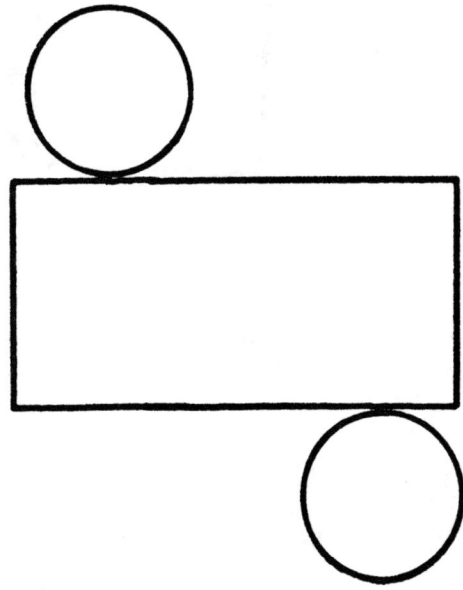

## VOLUME VENTURE

---
Skill: *Finding the volume of space figures*
---

1. Prepare a Volume Center by providing large models of a sphere, a cone, a pyramid, a cube, a cylinder, and a rectangular prism. (You might include other prisms also.)

   Make the space figures from lightweight cardboard. A beach ball could be used for the sphere.

2. On each figure, write the formula for finding the volume. Then, hang the space figures from the ceiling. Suspend them at a level low enough that students may examine them and read the formulas easily.

3. Fill a container with various objects that represent the figures. Provide rulers, measuring tapes, meter sticks. The objects might include:

   | pencils | Tootsie Rolls | tin cans |
   | megaphone | tennis ball | glass prism |
   | golf ball | small boxes | cylindrical glasses |
   | marbles | straws | ping pong ball |
   | dice | wood pieces | ice cream cone |

4. Direct students to find the volumes of the objects in the container and to keep a record of their measurements. Encourage them to find and measure other cones, cylinders, spheres, cubes, pyramids, and prisms in the classroom.

185

## A GALLON IS A GALLON IS A GALLON

**Skill:** *Using the standard units for liquid measure: liters, cups, pints, quarts, gallons*

1. Provide containers that will contain exactly:

   1 liter     1 pint     1 gallon
   1 cup      1 quart

2. Supply, or ask students to bring, assorted bottles and jars. Try to find a few that will hold precisely a liter, half liter, cup, pint, quart, or gallon. Label each container with a letter or color code.

3. Display the bottles. Ask students to estimate which ones might hold a gallon, or a liter, etc. Then, let them test to see how closely they guessed.

   ALSO

   Fill the quart container. Pour the water into different containers to see how full it can fill them.

4. Give the students directions for activities which will enable them to experiment with the different units of liquid measurement. For example:

   *Which container is almost 2.5 liters?*
   *Can Bottle C hold 2 quarts, 1 pint?*
   *How much water does Jar H hold?*
   *Which unit of measurement will you use to measure D?*
   *Will J and B hold 2 gallons?*
   *Give the volume of each container.*
   *Which container will hold exactly 3 cups?*

## FILL 'EM UP!

---
Skill: *Measuring liquid capacity*

---

1. Fill an old suitcase with beans, rice, or unpopped popcorn.

2. Provide containers of these sizes: liter, cup, pint, quart, half gallon, gallon. Label each one with its capacity.

3. Include some other containers of varying capacity. Label these with letters.

4. In the lid of the suitcase, paste a list of several questions and activities which give students opportunities to investigate, compare, and use the different units. For example:

       2 gal. = _____ qt.     *Which is more?*  *2 pts. or 1 liter?*
       3 qt. = _____ cups                       *10 liters or 3 gal.?*
       1 qt. = _____ pt.
       _____ gal. = 12 pts.

*Container A holds* _____ *gallons.*
*Container A holds* _____ *liters.*
*Does Container C hold closer to 7 pints or 5 liters?*

LOLLIPOPS I CAN LICK

---
Skill: *Choosing a suitable unit for a measurement task*
---

1. Cut several four-inch (10 cm.) circles of many colors. Make them into lollipops by stapling them to popsicle or paste sticks.

2. Number each lollipop and write a measuring task on it. If the task is one that could be done in the classroom, mark it with a star. Here are some ideas for tasks that would involve some kind of measuring:

    *frosting a cake*
    *choosing a box to mail a present*
    *putting air into a football*
    *buying carpeting*
    *buying a belt*
    *taking your temperature*
    *wrapping a present*
    *laying sod in your front yard*
    *filling a swimming pool*
    *deciding if the lake is deep enough for diving*
    *throwing a softball*
    *painting your bedroom*
    *pole vaulting*
    *drinking a milkshake*
    *dividing up a pizza*
    *finding the distance to Disneyland*

3. Place all the lollipops in a tall can. Directions can be glued to the can.

4. Duplicate a record sheet where students can write the unit they have chosen for each task. If the lollipop is starred, they must also do the measuring.

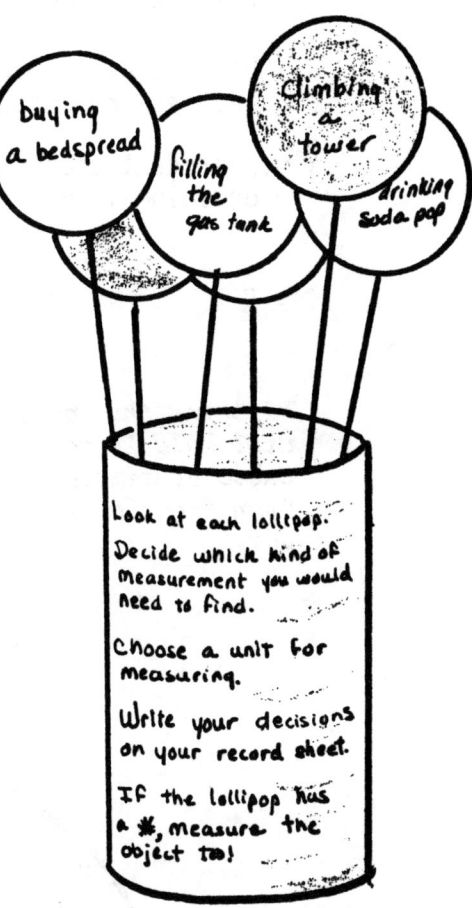

| Number | I would need to find - | Unit I'd choose | Measurement ✶ ones only |
|---|---|---|---|
| 1 | volume | cubic centimeter | |
| 2 | | | |

## JUDGE THE JUNK

---
**Skill:** *Estimating and comparing measurements*
---

1. Create a Judge the Junk Center. Fill a garbage can or large box with a great assortment of objects. Students will want to contribute their "junk" too!

2. In the center, provide: rulers, meter sticks, yardsticks, measuring tapes, liquid measuring units (liter, cup, pint, quart), beans, scales, centimeter cubes, pencils, paper, yarn.

3. Have students make personal notebooks for recording their estimates and measurements.

4. A student chooses two items from the box, compares their size, weight, or volume, and writes in his notebook an estimate for each.

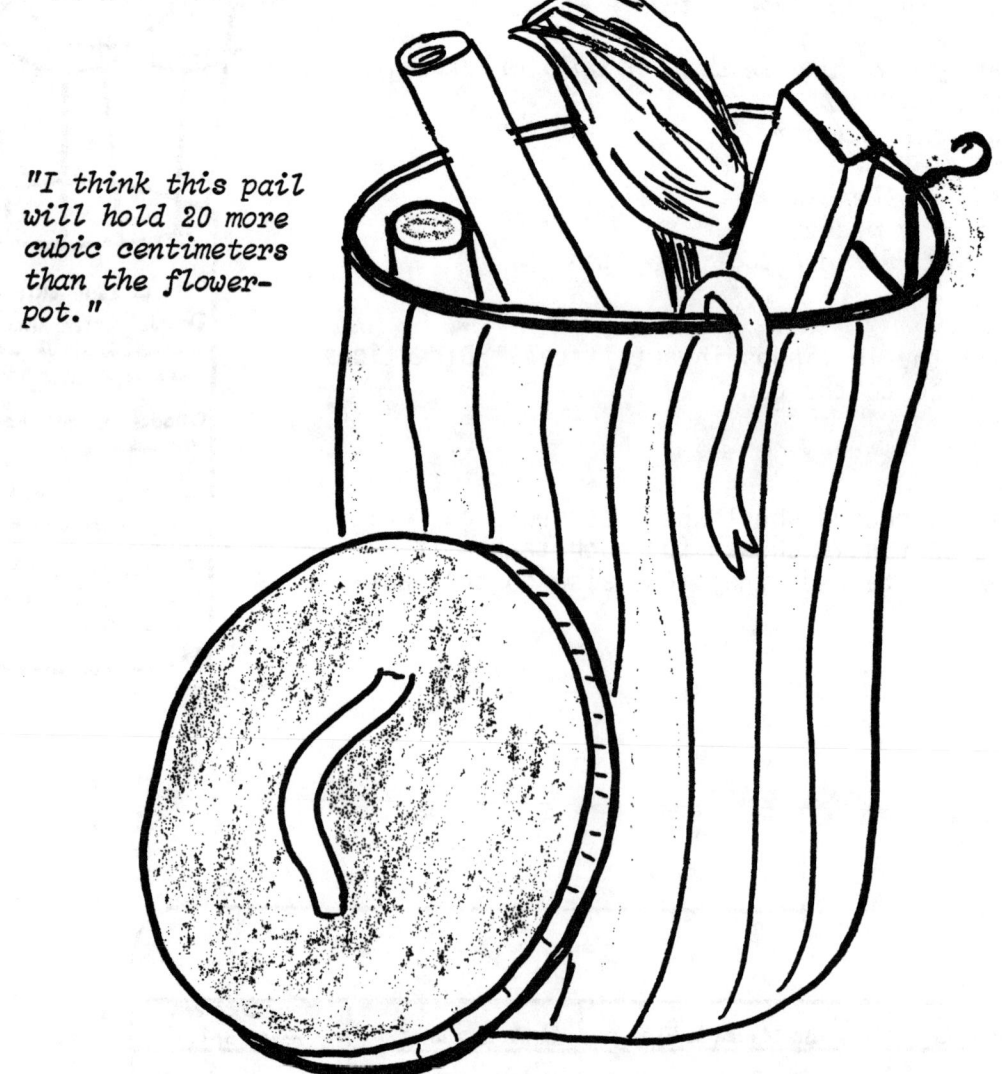

*"I think this pail will hold 20 more cubic centimeters than the flowerpot."*

5. Then, he makes the measurement and records his findings.

189

GEOMETRY

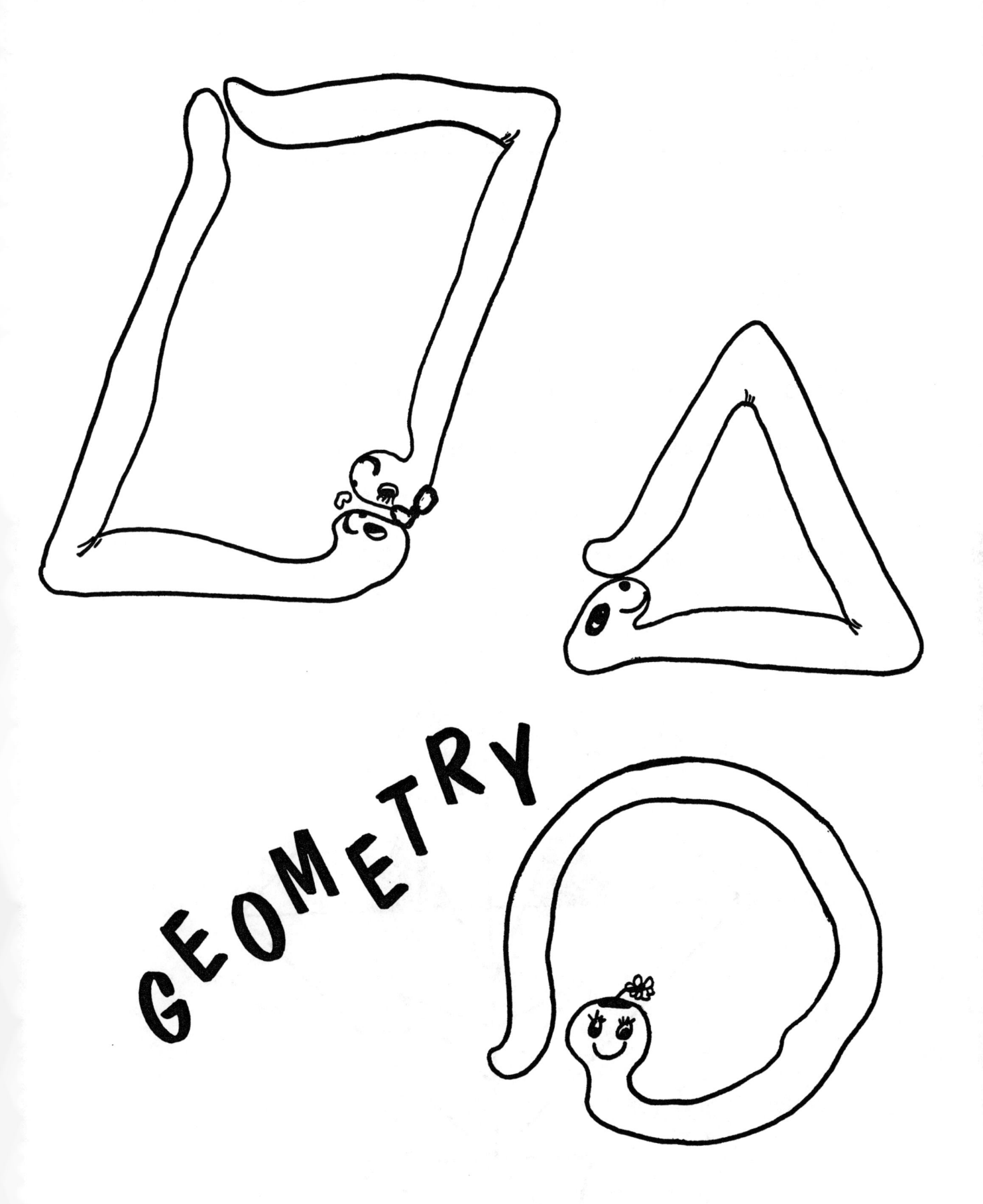

## COUNT AND COLOR

Skill: *Recognizing triangles and rectangles*

1. Give the students copies of the design below.

2. Ask them to count the triangles and color the design. Let them also try the one on the following page.

3. Have each student make a picture or design in which he has included many **triangles, rectangles** (perhaps squares or circles, too).

4. Students may trade designs and count the shapes they find.

*In Figure A there are 21 triangles.*

*In Figure B, there are 45 rectangles.*

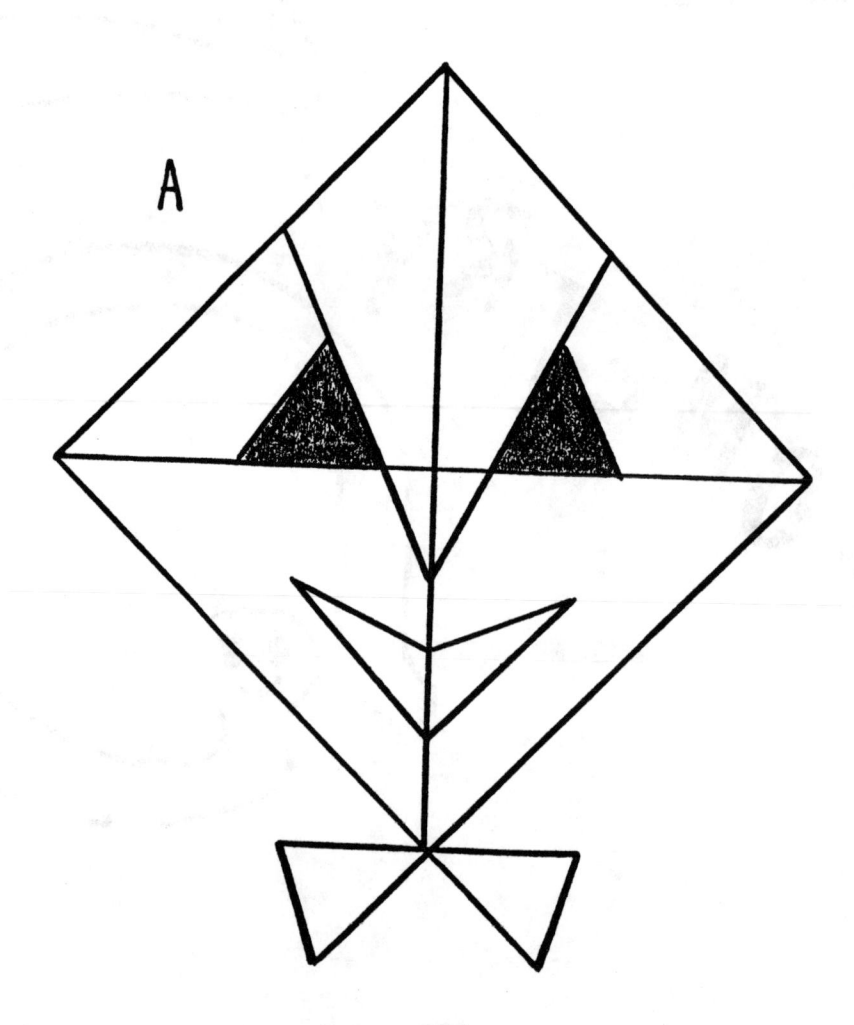

A

Name _____ *

B

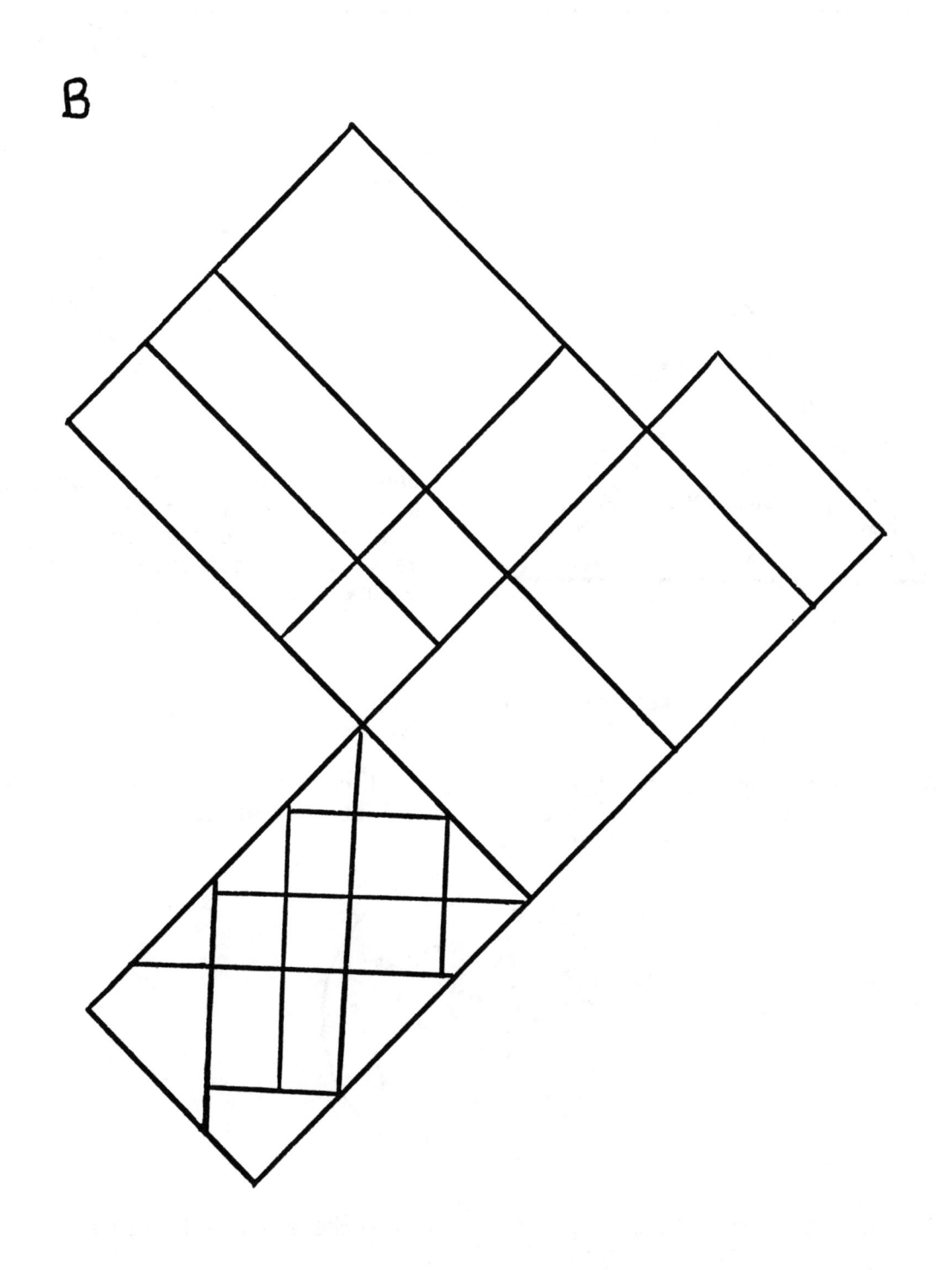

How many rectangles? _____

# TRIPLE INVENTORY

**Skill:** *Identifying circles, triangles and rectangles*

1. Cut enough circles (six-inch diameter), triangles (six-inch base), and rectangles (4 x 6 inches) so that every student can have one of each. Ask them to label each figure.

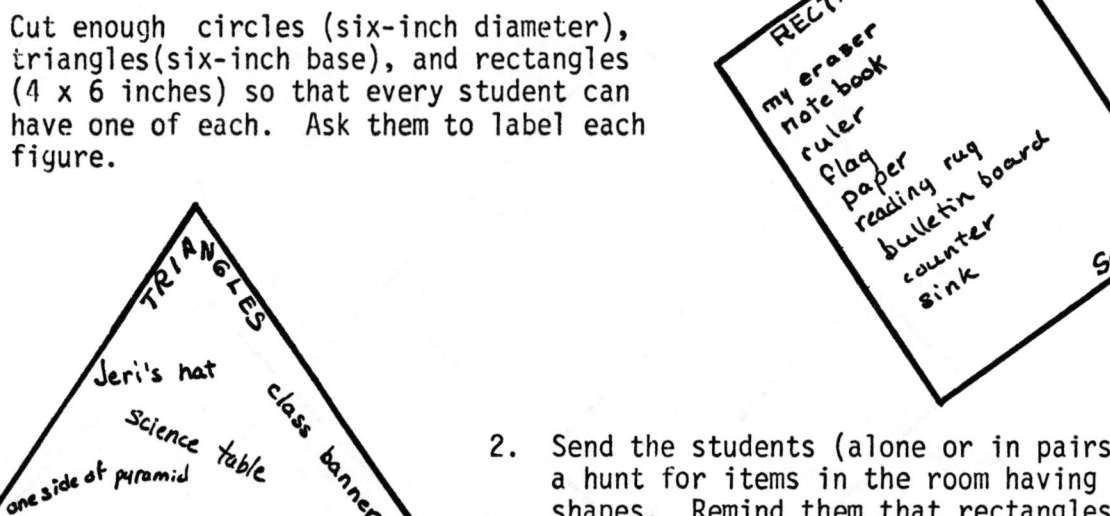

   *RECTANGLES*
   my eraser
   notebook
   ruler
   flag
   paper
   reading rug
   bulletin board
   counter
   sink
   Sam

   *TRIANGLES*
   Jeri's hat
   class banner
   Science table
   one side of pyramid
   our new flashcards
   Annie

2. Send the students (alone or in pairs) on a hunt for items in the room having those shapes. Remind them that rectangles may vary quite a bit from the one they have before them.

3. Each student writes his own list on his circle, triangle, or rectangle.

4. Have someone cut the three figures in larger sizes from poster board. A secretary may combine all the students' lists and write the items on these large figures, making a master list.

   *CIRCLES*
   lens on my glasses
   thumbtack
   hole in the pencil sharpener
   doorknob
   mole on John's arm
   my ring
   top of thermos
   Jodi

5. Hang the three large figures in a place where students can easily get to them for making any additions they find.

ANYBODY HOME?

---
Skill: *Identifying closed curves and their interior and exterior regions*
---

1. Draw several closed curves. Make some of them simple and some, complex.

2. Display the curves on a wall or bulletin board. Place several ants near or inside each curve. Give each ant a number.

3. Tell the students that each figure is an ant home and that they must decide which ants are <u>inside</u> their homes and which are not.

4. Have a large <u>answer</u> ant available for self-checking.

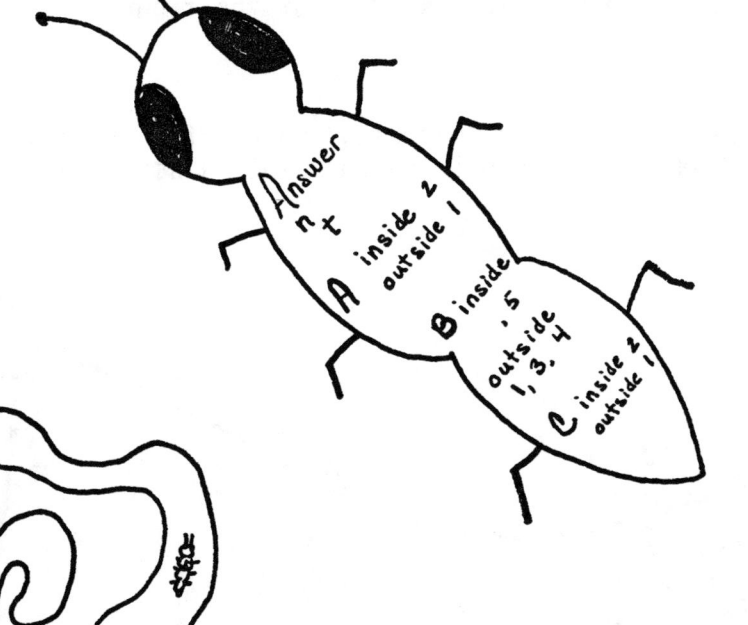

Answer Ant

A inside 2 outside 1

B inside 5 outside 1, 3, 4

C inside 2 outside 1

194

# CURVACEOUS . . . GRACIOUS!

---
Skill: *Recognizing and making simple and complex closed curves*

---

1. Provide two large sheets of white construction paper (12 x 18 inches) for each student.

2. Have students draw one large simple closed curve and one large complex closed curve.

3. Ask students to trade both papers with someone, and

    a. make the simple closed curve into a picture of a person, object, or scene;

    b. make the complex closed curve into a colorful design, using a different pattern in each section of the figure.

4. Display the curve art by "wallpapering" your math corner with the pictures, hanging them from a wire or putting them into a scrapbook.

# TRY ANGLES

---
Skill: *Identifying various kinds of triangles*
---

1. Make five scrolls as shown below, using paper about 20 x 30 inches. Four scrolls define and picture classes of triangles. The fifth shows examples of all kinds of triangles.

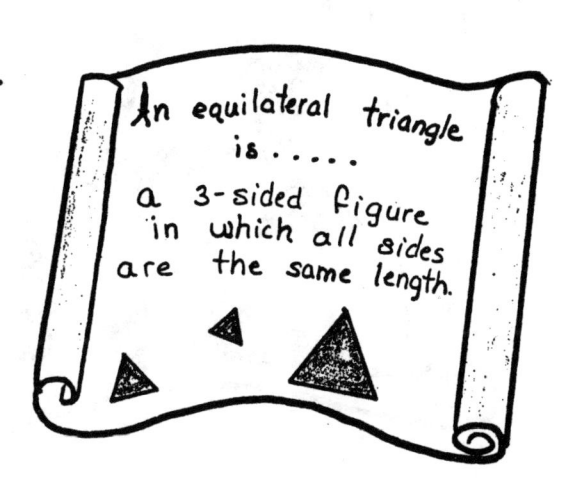

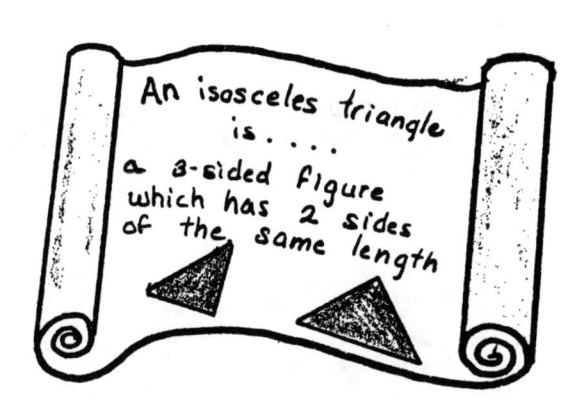

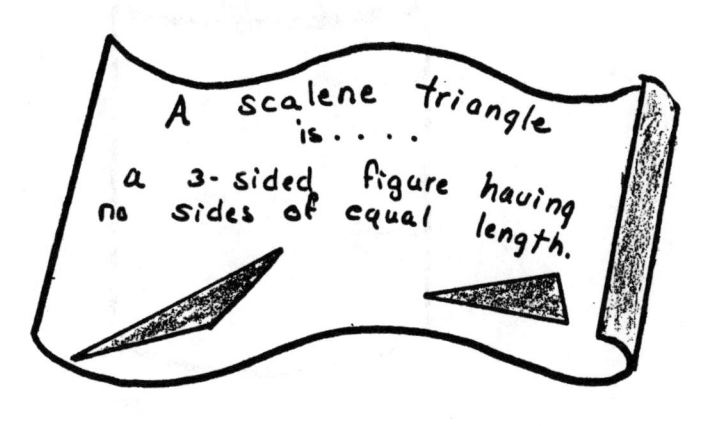

   *An equilateral triangle is..... a 3-sided figure in which all sides are the same length.*

   *A right triangle is...... a 3-sided figure that has one right angle.*

   *An isosceles triangle is.... a 3-sided figure which has 2 sides of the same length.*

   *A scalene triangle is.... a 3-sided figure having no sides of equal length.*

2. Label each scroll on the outside.

196

3. A student may take the ALL SORTS scroll and one of the other four. He then examines the triangles to find which ones fit the description of the kind he is studying.

4. Provide protractors and rulers for measuring-- and an answer key for checking!

WHO'S WHO IN QUADRILATERALS?

---
Skill: *Recognizing properties of different quadrilaterals*
---

1. Form about ten different quadrilaterals from thick yarn dipped in a mixture of white glue and water (glue:water--2:1). Lay the shapes on waxed paper, overlapping ends.

2. When they are dry, hang the quadrilaterals underneath a table. Label each shape in some way--by color or by attaching a letter.

3. Under the table, place a container for pencils and protractors.

4. A student may crawl under the table to examine the properties of the quadrilaterals and record his observations on a data table such as the one on the next page.

# Who Am I?

**Directions:** Look at each figure carefully. If it fits into one or more of the five classifications, write *yes* in that box.

| Figure | Rectangle? 2 pairs of parallel sides 4 right angles | Rhombus? 4 congruent sides | Square? 4 congruent sides 4 right angles | Parallelogram? 2 parallel sides (or more) | Trapezoid? only 2 parallel sides |
|---|---|---|---|---|---|
| A | | | | | |
| B | | | | | |
| C | | | | | |
| D | | | | | |
| E | | | | | |
| F | | | | | |
| G | | | | | |
| H | | | | | |
| I | | | | | |

Name _____

## GO GEOMAT STYLE

---
Skill: *Recognizing and defining plane geometric figures*

---

1. Make a geomat by using permanent marker to draw many geometric figures on a window shade.

2. Prepare a set of figure cards. On each card, draw a geometric figure and label it. On the back of the card, write a description or definition for the figure. The next page suggest some figures you might include.

3. On the bottom of the shade, write the instructions to the students. Direct them to read each card, then to search the geomat for examples of that figure.

200

4. Students then use the mat and the cards for learning to identify plane geometric figures.

Variations:
1. You may make your geomat as simple or complex as is fitting to your students' abilities by varying the number and kinds of figures you include.

2. Color code the cards and figures so that they can be grouped according to difficulty.

---

**FIGURES TO INCLUDE ON YOUR GEOMAT . . .**

line segment

point

angle

intersecting line segments

vertex

parallel line segments

acute angle

right angle

obtuse angle

triangle

right triangle

scalene triangle

isosceles triangle

circle

equilateral triangle

arc

diameter

radius

quadrilateral

rectangle

parallelogram

# SORT SPORT

---
Skill: *Classifying plane geometric figures*
---

1. Tie or tape together 17 cylindrical containers (oatmeal boxes or cans with one end removed). The containers might be painted or covered with colorful adhesive paper before they are joined.

2. Fill the top container with pictures of angles, triangles, quadrilaterals, and other polygons drawn on small cards.

3. Label the containers as shown here.

4. Individual students may classify the figures into the four main groups.

5. For further development of classification skills, ask the student to separate each of the four groups into the sub-classes shown. Be sure to provide an answer key.

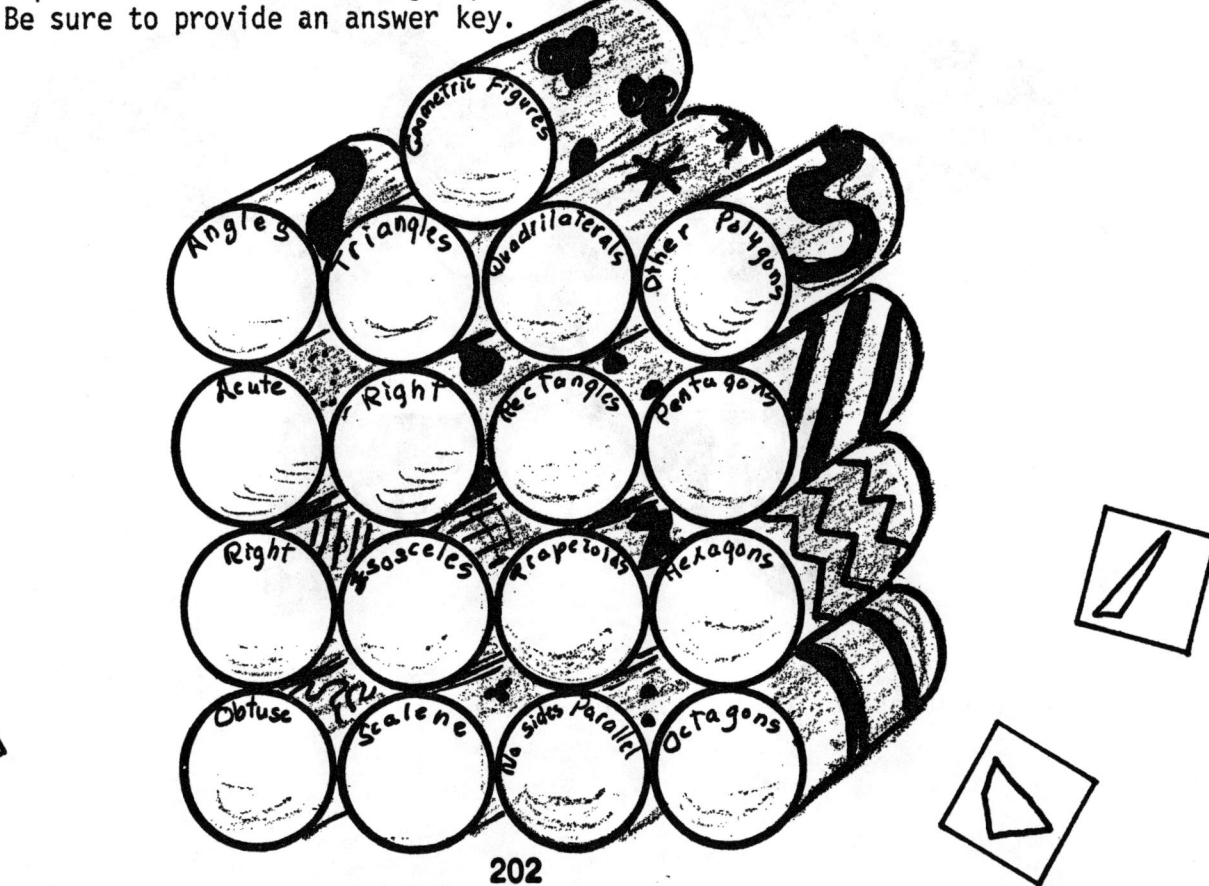

KALEIDOSCOPE

---
Skill: *Using a compass to draw circles*

---

1. Give students large sheets of newsprint on which to practice drawing circles. (Hint: *A workbook under the paper gives the compass point a firm grip, while it may slip on a hard surface.)

   *Tell students to turn the paper, not the compass. That way, the fingers can firmly grip the compass and it won't be likely to slip.

2. Show some ideas for making designs from circles and parts of circles. Try: overlapping circles, arcs, half circles, circles within circles, parts of circles within squares, etc.

3. Provide additional paper for drawing designs. Encourage students to experiment before settling on a design.

4. Let students color the designs with marker, crayon, ink, or paint.

   Variation: Older students will like working alone or in pairs on large, poster-sized designs. Fluorescent paint looks especially great on big geometric posters.

Scramble some of the geometric words:

| ribsecot | niratleg | lehnoydop | nepal |
| galen | cafe | beas | rac |
| reevxt | mayrdip | liequalrate | miperteer |
| gootnac | diusar | noce | plif |

Answers:

| *bisector* | *triangle* | *polyhedron* | *plane* |
| *angle* | *face* | *base* | *arc* |
| *vertex* | *pyramid* | *equilateral* | *perimeter* |
| *octagon* | *radius* | *cone* | *flip* |

## CIRCE-FILE

---
Skill: *Identifying and drawing circles and parts of circles*
---

1. Supply each student with one or more nine-inch tagboard squares.

2. Ask students to use compasses and rulers to draw a picture which includes at least one circle. Encourage them to add parts of circles (arcs, diameters, radii, chords, etc.) to their drawings.

3. When the squares are completed and colored, number them (from simple to complex) and file them in a box.

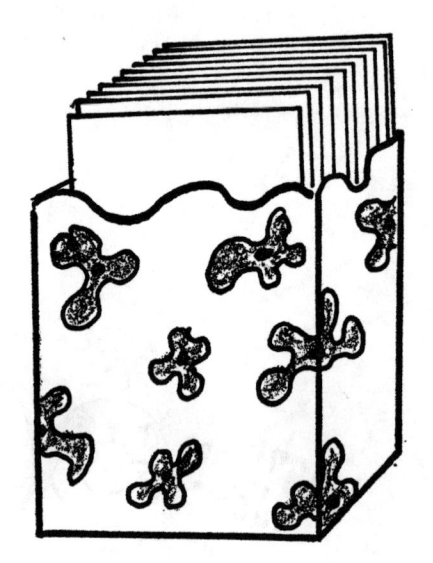

4. Add a card to the box which directs students to

    count the circles on each card

    OR

    identify radii, diameters, chords, and arcs

    OR

    measure the circumferences.

205

## GEO-BINGO

---
Skill: *Recognizing plane geometric figures*
---

1. Give each student a copy of each of the two following pages.

2. Ask students to cut apart the squares on Sheet A and paste them in any arrangement on the spaces of Sheet B.

3. Make a set of small calling cards for the Bingo game. On each card, write a label for one of the figures shown:

| | | |
|---|---|---|
| *acute angle* | *chord* | *obtuse angle* |
| *triangle* | *tangent* | *intersecting line segment* |
| *circle* | *right triangle* | *diameter* |
| *ray* | *rectangle* | *perpendicular line segment* |
| *square* | *arc* | *closed curve* |
| *right angle* | *curve* | *rhombus* |
| *line segment* | *quadrilateral* | *parallelogram* |
| *trapezoid* | *obtuse angle* | *pentagon* |
| *hexagon* | *ellipse* | *bisector* |

4. Students will need to cut small pieces of paper for markers.

5. Each student uses the Bingo card he has made as the class or a group plays Geo-Bingo.

Variation: Make the Bingo game easier or more difficult by varying the figures used.

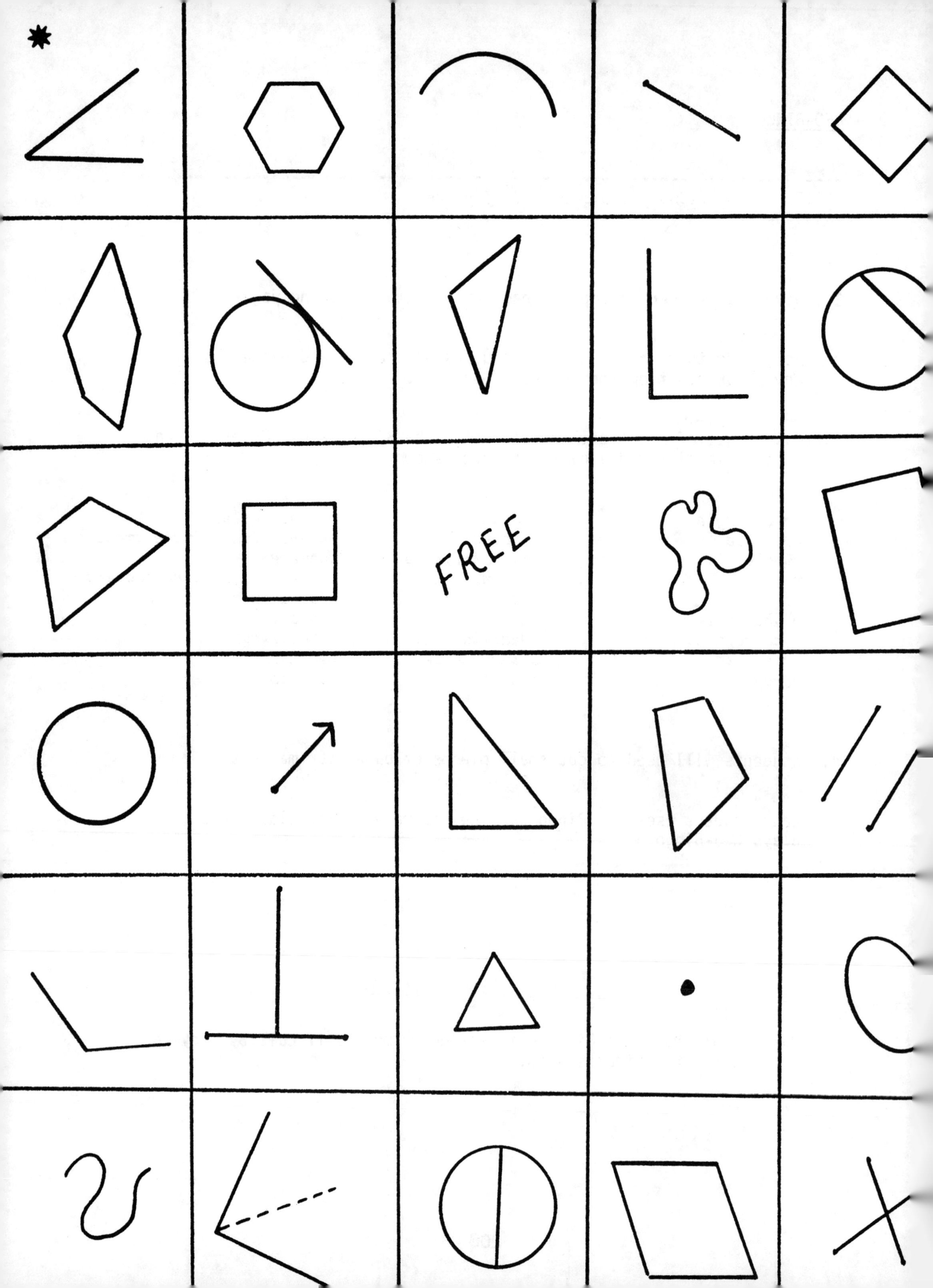
FREE

※ B

## STITCH-WITCHERY

---
Skill: *Making various kinds of plane geometric figures*
---

1. Provide:   large needles
              yarn
              styrofoam meat trays or scraps of thin cardboard

2. Ask students to stitch and label geometric figures.

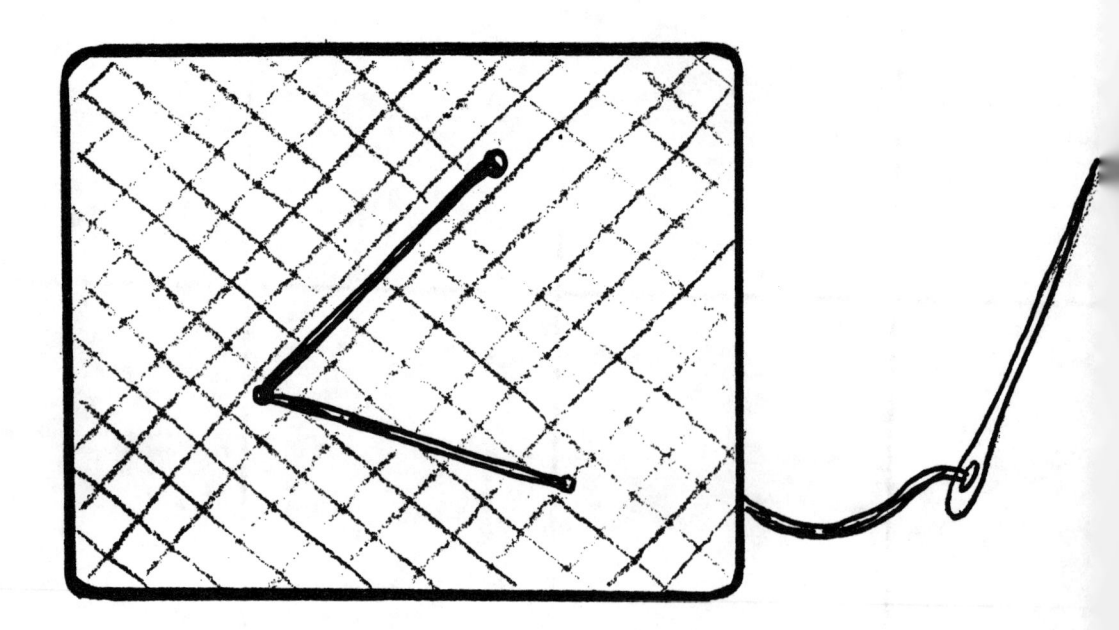

Variation:   Have students stitch figures on burlap squares and sew the squares together for a wall hanging.

## TRY A TANGRAM

---
Skill: *Making plane geometric figures*
---

1. Cut several sets of tangram pieces (patterns on the next page) or have each student cut a set for himself.

2. Use an envelope to keep the pieces of each set together.

3. Ask students to use all or some of the tangram pieces to make the following figures. (See illustration below.)

4. Encourage them to use the pieces to invent their own shapes. Have them trace around each shape they make and save it so that others may try to fit pieces into their outlines. There are hundreds of shapes that can be created!

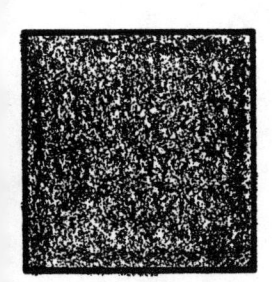

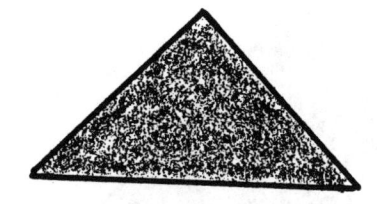

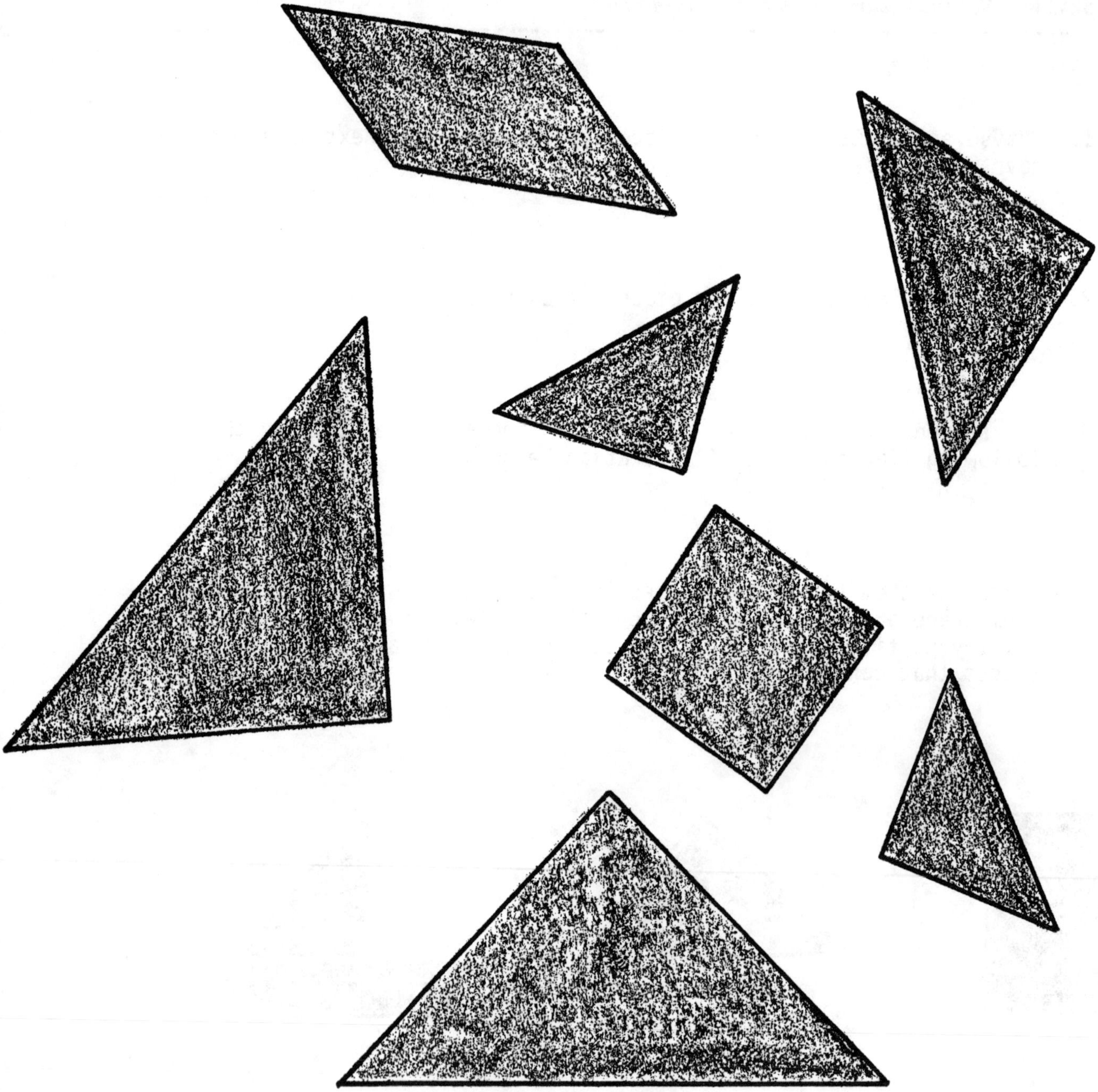

## GO GEOBOARDS OR BUST!

**Skill:** *Making plane geometric figures of various sizes*

1. Create a corner where students may find and use the supplies they need for making figures on geoboards.

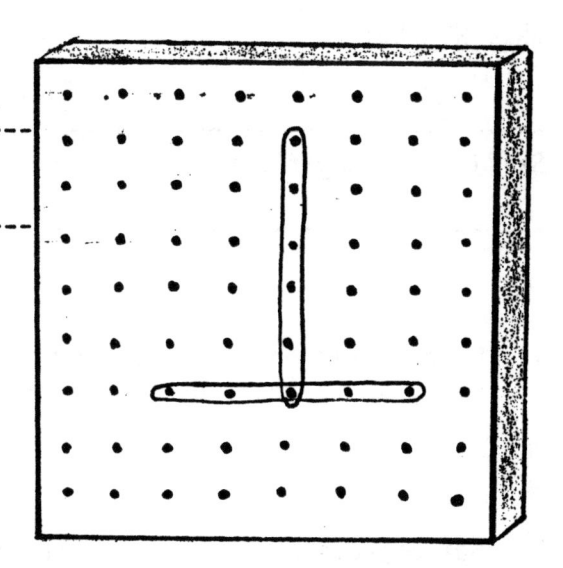

2. Provide geoboards, rubber bands, pencils, and activity cards with directions such as:

    a. *Show two parallel line segments.*

    b. *Show two perpendicular segments.*

    c. *Show a line segment five units long.*

    d. *Show an acute angle.*

    e. *Show two congruent angles.*

    f. *Make the smallest possible triangle. Now, make one twice as big.*

    g. *Show a trapezoid with a base three units long.*

    h. *Make a hexagon with each side two units long.*

3. Have some "dot paper" available--this enables students to draw the figures they have shown on their geoboards.

## GRID WORKOUT

**Skill:** *Finding the perimeters and areas of polygons*

1. Make a grid on a transparency. Show it, using an overhead projector.

2. Set objects on the grid. Ask the students to figure the perimeter of the object or to tell the area of the grid covered by the shape.

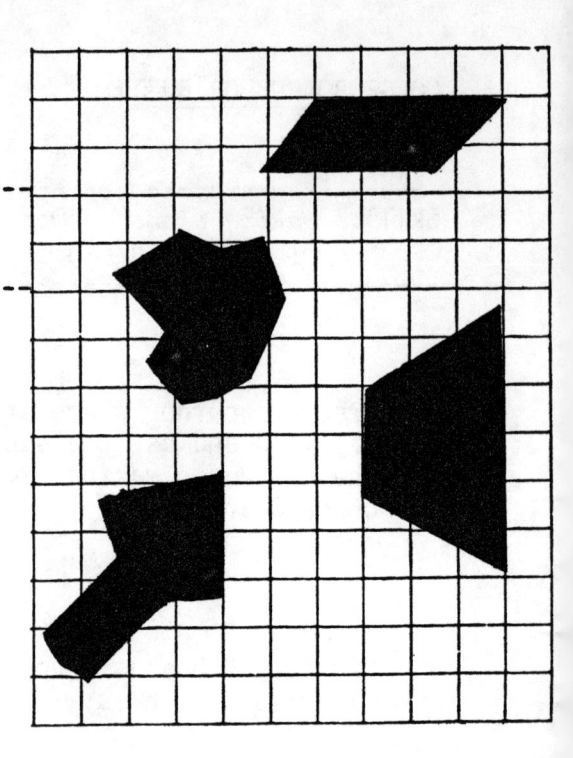

3. Let students repeat the activity with other shapes and objects in the room.

4. Distribute copies of the grid on the next page (or create a similar one, fitting the abilities of your own students).

5. Ask students to write the perimeter and/or area of each polygon.

6. As a follow-up activity, provide plain grids for students and let them cover portions of the grids with polygons they have cut or found.

Name

MATCH GAME

---
Skill: *Recognizing the properties of polygons*
---

1. Make three sets of cards from construction paper. The next page describes the cards and supplies some ideas.

2. Glue a piece of felt of flannel to the back of each card.

3. Supply a large flannel board or a piece of flannel cloth (one yard square) and crayons.

4. A student matches the names with the description cards. For each pair, he uses a blank card to draw an example of the polygon.

215

# DESCRIPTIONS OF GEOMETRIC FIGURES

SET I   (3 x 5 inches)

   a. Two of my three sides are congruent.
   b. My four sides might be of any length.
   c. I have six sides.
   d. All four of my sides are parallel but I don't always have right angles.
   e. All three of my sides are congruent.
   f. I am a quadrilateral with only two parallel sides.
   g. I have four congruent sides and four right angles.
   h. I have three sides and one right angle.
   i. I have four parallel sides and four right angles.
   j. I have eight sides.
   k. My four sides are congruent but I don't necessarily have right angles.
   l. No two of my three sides are congruent.


SET II   (1 x 5 inches)   Names of Polygons

   right triangle              trapezoid
   equilateral triangle        square
   rectangle                   rhombus
   parallelogram               hexagon
   pentagon                    octagon
   isosceles triangle          quadrilateral
   scalene triangle

SET III   (5 x 5 inches)   Blank cards (cover with clear adhesive)

*bright idea!*

Collect scraps of fabric from which students may cut polygons. Let them sew together the figures to create a geometric quilt.

## THESE ANGLES HAVE CLASS!

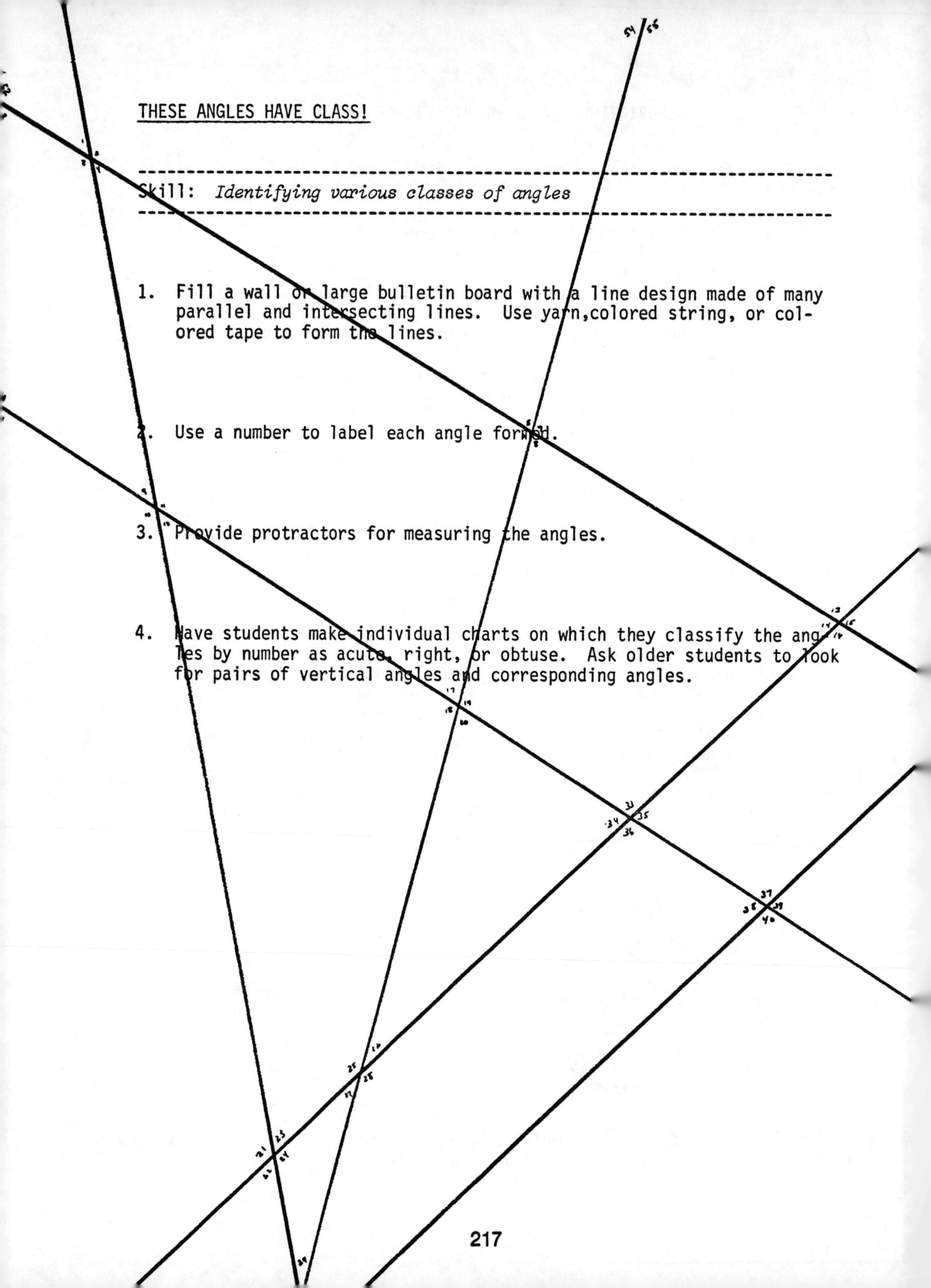

---
Skill: *Identifying various classes of angles*
---

1. Fill a wall or large bulletin board with a line design made of many parallel and intersecting lines. Use yarn, colored string, or colored tape to form the lines.

2. Use a number to label each angle formed.

3. Provide protractors for measuring the angles.

4. Have students make individual charts on which they classify the angles by number as acute, right, or obtuse. Ask older students to look for pairs of vertical angles and corresponding angles.

CHECK YOUR ANGLE

---
Skill: *Finding and measuring angles in plane figures*
---

1. Turn an origami art project into a geometry exercise! You'll need to supply:

    thin colored paper in assorted sizes (wrapping paper, origami art paper, or colored newsprint)

    pencils or crayons

    protractors

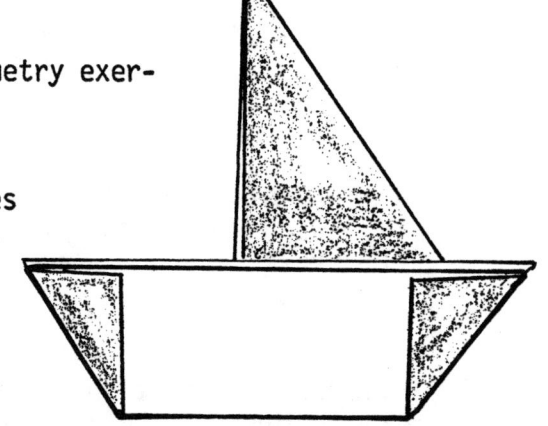

2. Let students experiment with paper folding until they have made at least one shape.

3. Then ask them to <u>unfold</u> their papers completely and examine the angles formed by the folds.

4. Have students measure and label each angle they find. Encourage them to look for pairs of congruent angles, vertical angles, and corresponding angles.

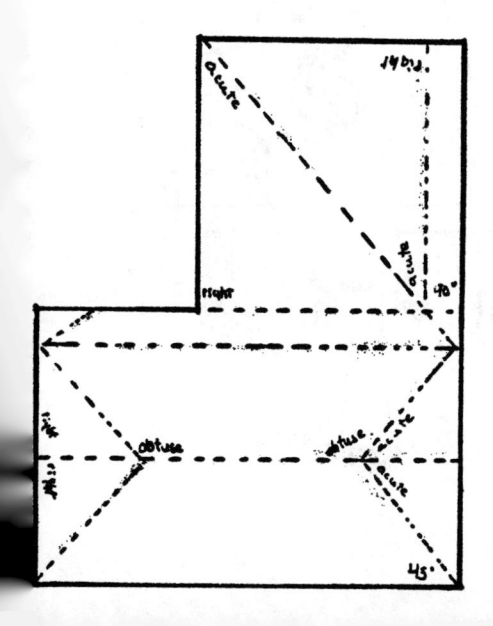

218

## WHAT'S DOT?

---
Skill: *Recognizing and constructing congruent plane figures*
---

1. Draw dots at one-inch intervals on a plain piece of chart paper or tagboard (24 x 36 inches or larger).

2. Draw several plane figures on the chart, making some congruent and labeling each figure with a letter. Leave some spaces where students may draw figures.

3. Hang the dot chart on an easel. Cover it with clear plastic adhesive or hang a piece of acetate over the chart. Supply protractors, compasses, rulers, crayons or water soluble markers and a cloth.

4. In one corner of the easel, hang a list of instructions for students, including tasks such as:

   *Which figure is ≅ to C?*
   *Draw a figure ≅ to ∠X.*
   *Is D ≅ L?*
   *Is this true? △B ≇ △R*
   *Draw a line segment ≅ to $\overline{FG}$*

5. A student may work at the easel to answer the question and complete the tasks. When he is finished, he uses the cloth to wipe off any figures he has drawn.

6. Each student visiting the easel might be asked to add one question or task to the list.

# A BIT O' BISECTING

---
Skill: *Constructing bisectors of line segments and angles*
---

1. Reproduce the following two pages. Give copies to each student.

2. Students may follow the directions to learn and practice bisecting line segments and angles.

3. Make several pictures of objects representing line segments and angles. (See examples below.) Mount these on 9 x 9 inch cards and cover with clear plastic.

4. Provide compasses, water soluble markers, and a convenient spot for students to practice constructing more bisectors. Students may work on the cards and wipe off their markings when finished.

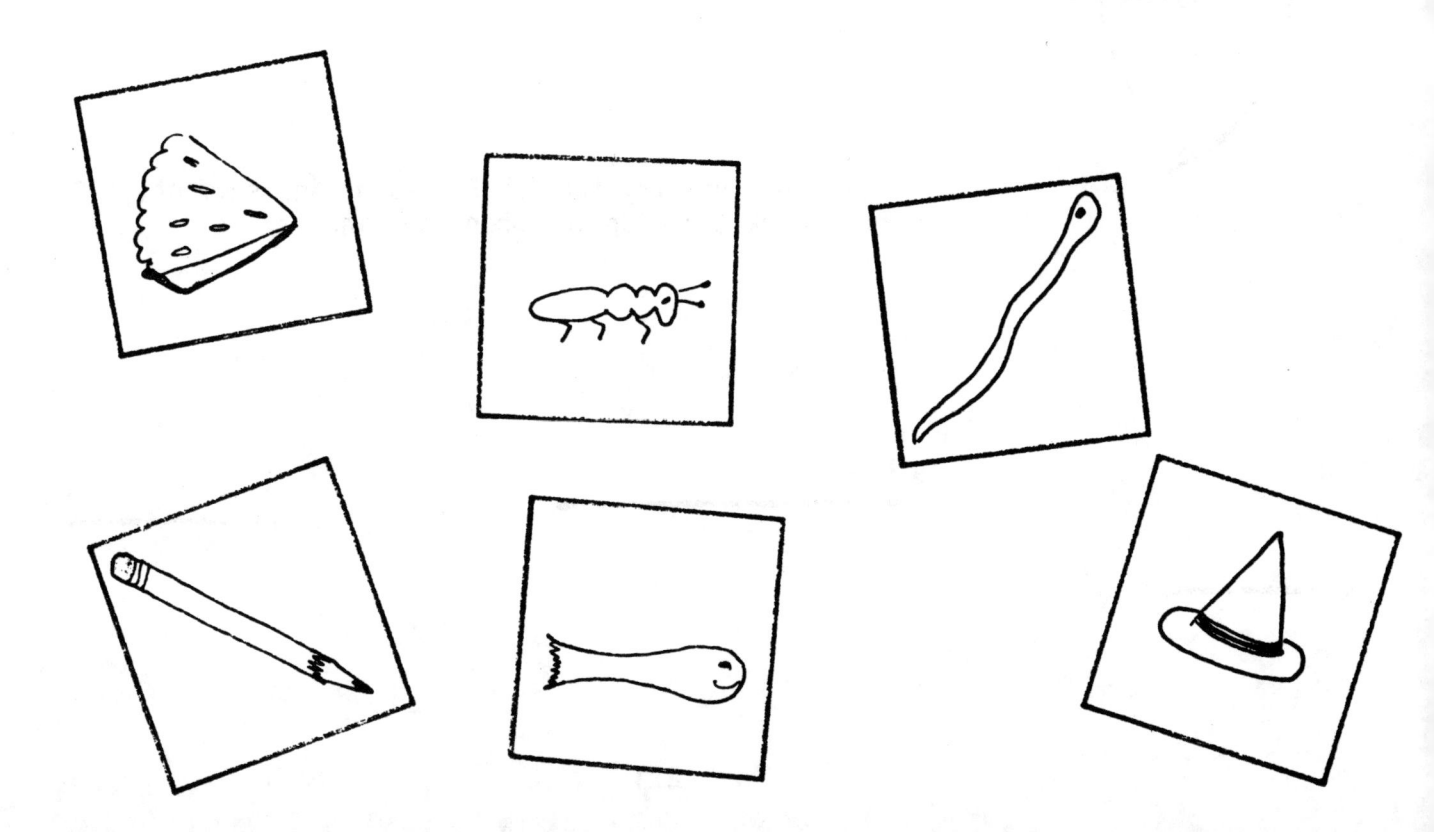

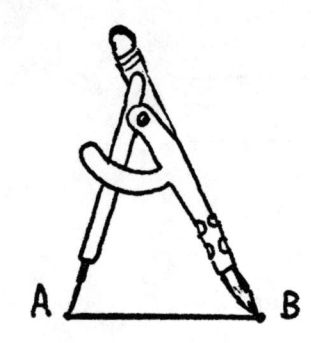

Name _____

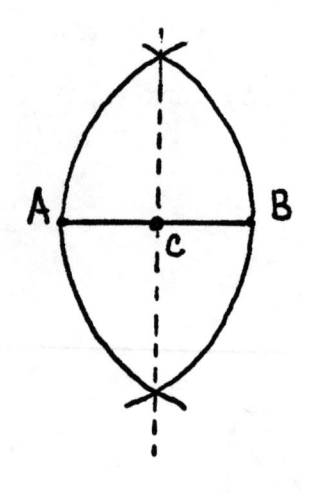

## HOW TO BISECT A LINE SEGMENT

1. Place the tip of your compass on A. Draw an arc through B.

2. Place the tip of your compass on B. Draw an arc through A.

3. Draw a line connecting the intersections of the arcs.

4. The point where this line meets line segment AB is called the midpoint of AB. Mark it C. You have bisected AB.

Now, you try these! For each line segment, mark the midpoint when you find it.

X———————Y   R————S

C——D

221

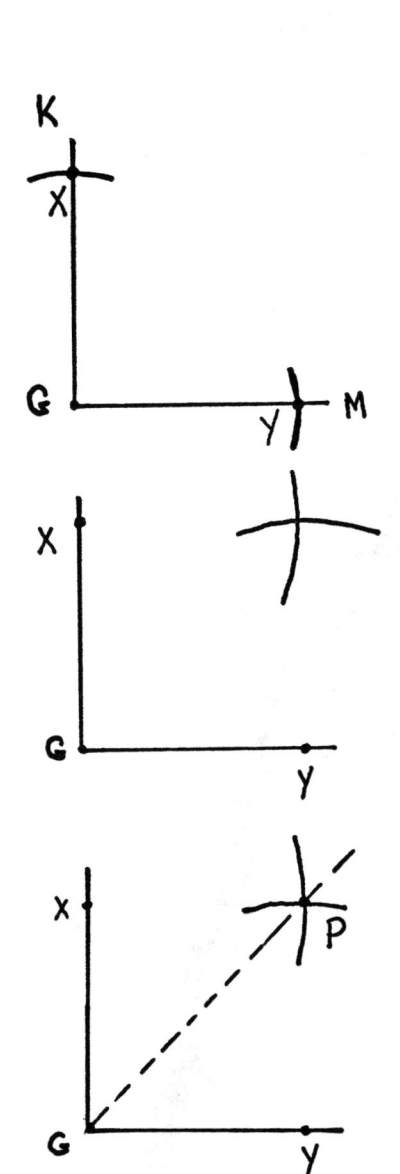

## HOW TO BISECT AN ANGLE

1. Place the tip of your compass on G. Draw an arc across GM.

2. Without changing the size of the compass opening, draw an arc through GK.

3. Mark points X and Y where the arcs intersect the line segments.

4. With the tip of your compass on X, draw an arc.

5. With the tip on Y, draw an arc that intersects the arc you just drew.

6. Draw a line from G through the intersection of the arcs. Label that point P.

7. Line segment GP is a bisector of Angle KGM.

You bisect these angles!

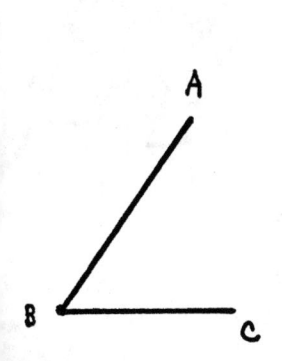

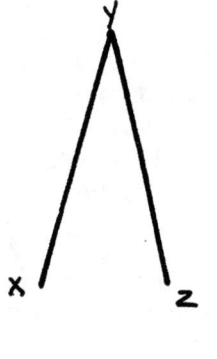

STICK OUT YOUR NECK--SPOT CHECK!

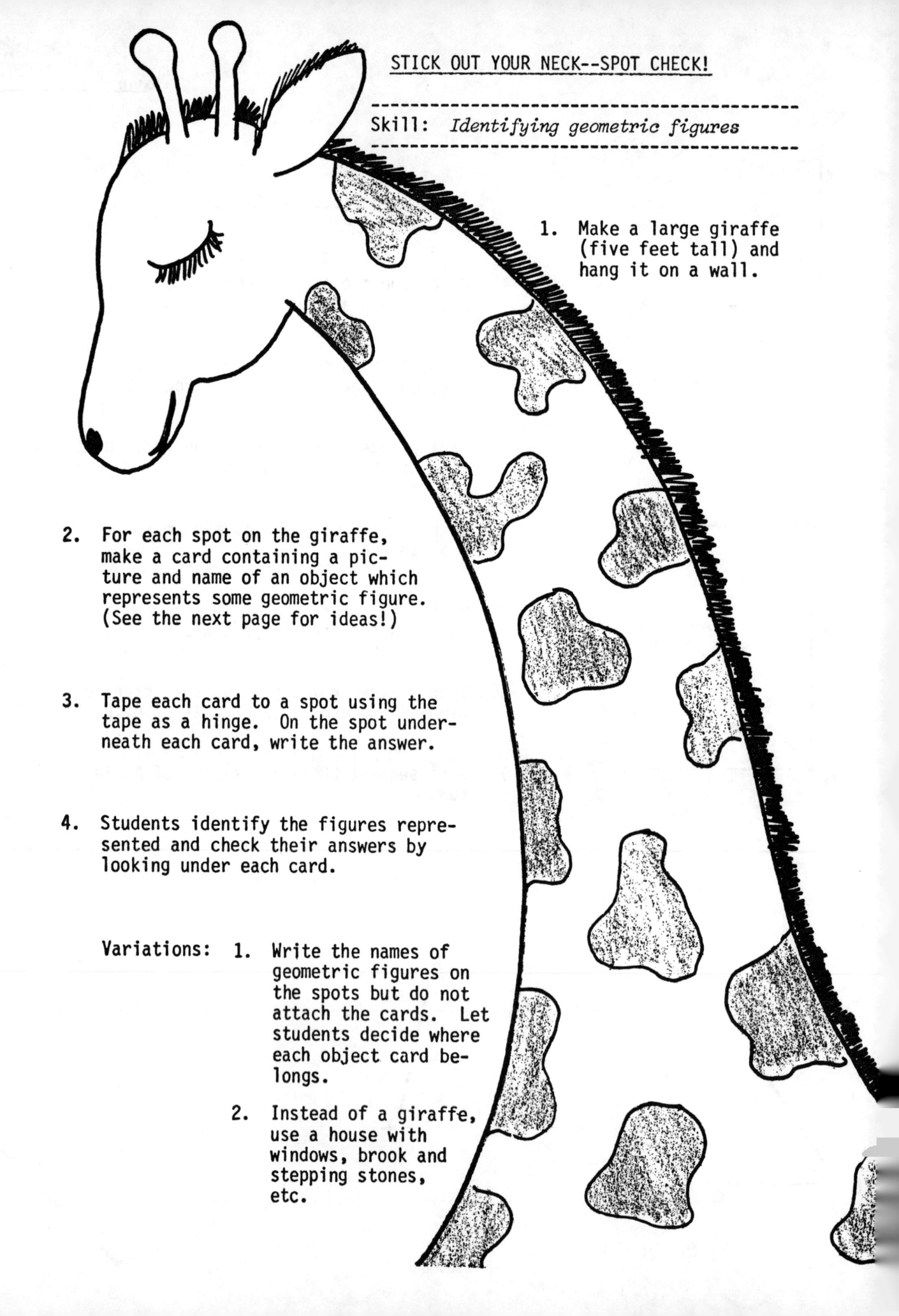

Skill: *Identifying geometric figures*

1. Make a large giraffe (five feet tall) and hang it on a wall.

2. For each spot on the giraffe, make a card containing a picture and name of an object which represents some geometric figure. (See the next page for ideas!)

3. Tape each card to a spot using the tape as a hinge. On the spot underneath each card, write the answer.

4. Students identify the figures represented and check their answers by looking under each card.

Variations:
1. Write the names of geometric figures on the spots but do not attach the cards. Let students decide where each object card belongs.

2. Instead of a giraffe, use a house with windows, brook and stepping stones, etc.

- ice cream cone
- box of cereal
- barrel
- aquarium
- dice
- a puddle
- party hat
- tent
- belly button
- rainbow
- button
- Stop Sign
- megaphone
- teepee
- snake
- Tootsie roll
- candle
- balloon
- rug
- baseball
- arrow
- ice cube

# STRAW GEOMETRY

**Skill:** *Making geometric figures*

1. Since straws represent line segments, they provide a convenient, fun means for learning about geometric figures. They are easy to handle, store, and cut.

2. Provide straws for each student. Make longer straws for yourself by fitting two together.

3. Hold up your straws to show students each figure you wish to teach. Define the figure, and ask students to show it with their straws.

4. Supply scissors and masking tape (or soft clay) for making polygons and building polyhedra. Possibilities are unlimited, and your students will love straw-geometry!

Variation: For individual practice, give students a worksheet listing figures you'd like them to make with their straws. Leave a space on the sheet where they may draw what they have made.

225

# GEOMOBILES

---
Skill: *Identifying space figures*
---

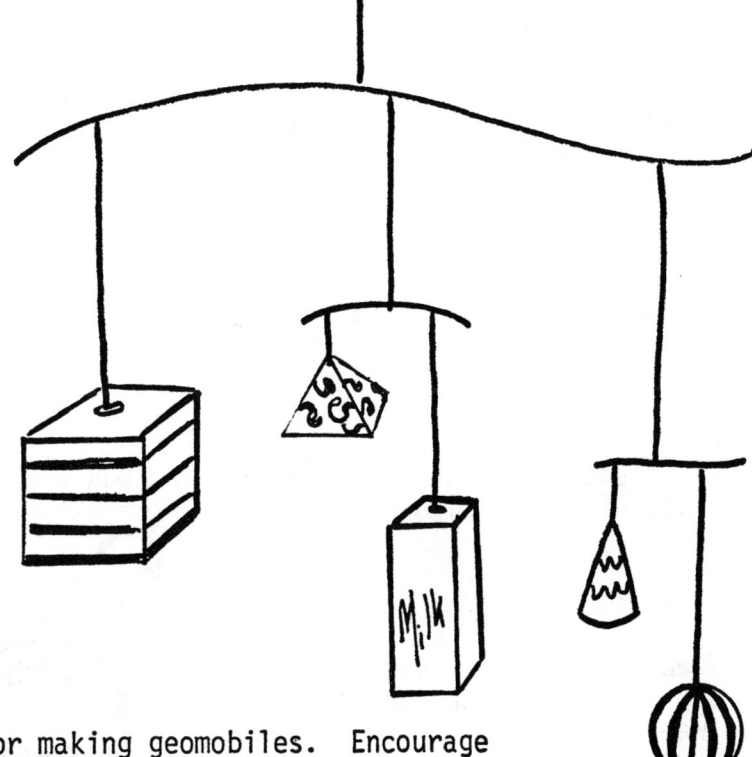

1. Provide materials for making geomobiles. Encourage students to bring materials from home:

   | | | |
   |---|---|---|
   | hangers | wooden dowels | |
   | string | yarn | |
   | small boxes | papier-mache | cardboard |
   | foam rubber | newspaper strips | fabric remnants |

2. Ask each student to make a mobile containing at least three space figures.

3. Hang the mobiles, labeling each with the artist's name.

4. Have students make and complete individual charts which list the names of the figures found on each mobile.

## SHAKE, RATTLE 'N ROLL

Skill: *Identifying space figures*

1. Obtain a set of wooden or plastic space figures. (See the Appendix for suppliers of math materials.)

2. Put one figure into a box and cover the box securely.

3. Let students shake, turn, or move the box and "listen" to the figure.

4. Then, students may ask you questions which can be answered Yes or No.

5. Set a limit of ten questions that may be asked as they try to identify the figure.

6. When one student properly identifies the contents of the box, he may choose another figure. Continue the game with him as the new leader.

## MULTI-DIP DIAGNOSIS

---
Skill: *Identifying the properties of space figures*
---

1. Display several space figures. Number each one.

2. Give students large pieces of construction paper. Ask them to draw a cone for each figure. In each cone, they may draw ice cream scoops to describe the properties of the different figures. They may label each cone with the name of the figure.

   Variation: Display several empty cones (numbered) on a bulletin board near the figures. Pin up an envelope containing "scoops" on which you have written properties of the figure. Let students build the ice cream cones by pinning each scoop on its corresponding cone.

(Faces / Edges / Vertices / Shape of base / Other Properties / Name of Figure)

HARD HAT AREA

---
Skill: *Constructing space figures*
---

1. Give students copies of the figures on the following pages. (Add other figures if you wish.)

2. Tell students to cut on the solid lines, fold on the dotted lines, and tape at each tab.

3. You may wish to have students

    --identify each figure before folding;

    --color and label figures before taping;

    --Count faces or measure surface area before folding.

Did you know that, for any polyhedron, the number of faces and vertices less the number of edges is 2?

$$(F + V) - E = 2$$

Have students count faces, vertices, edges--see if they can discover the relationship!

230

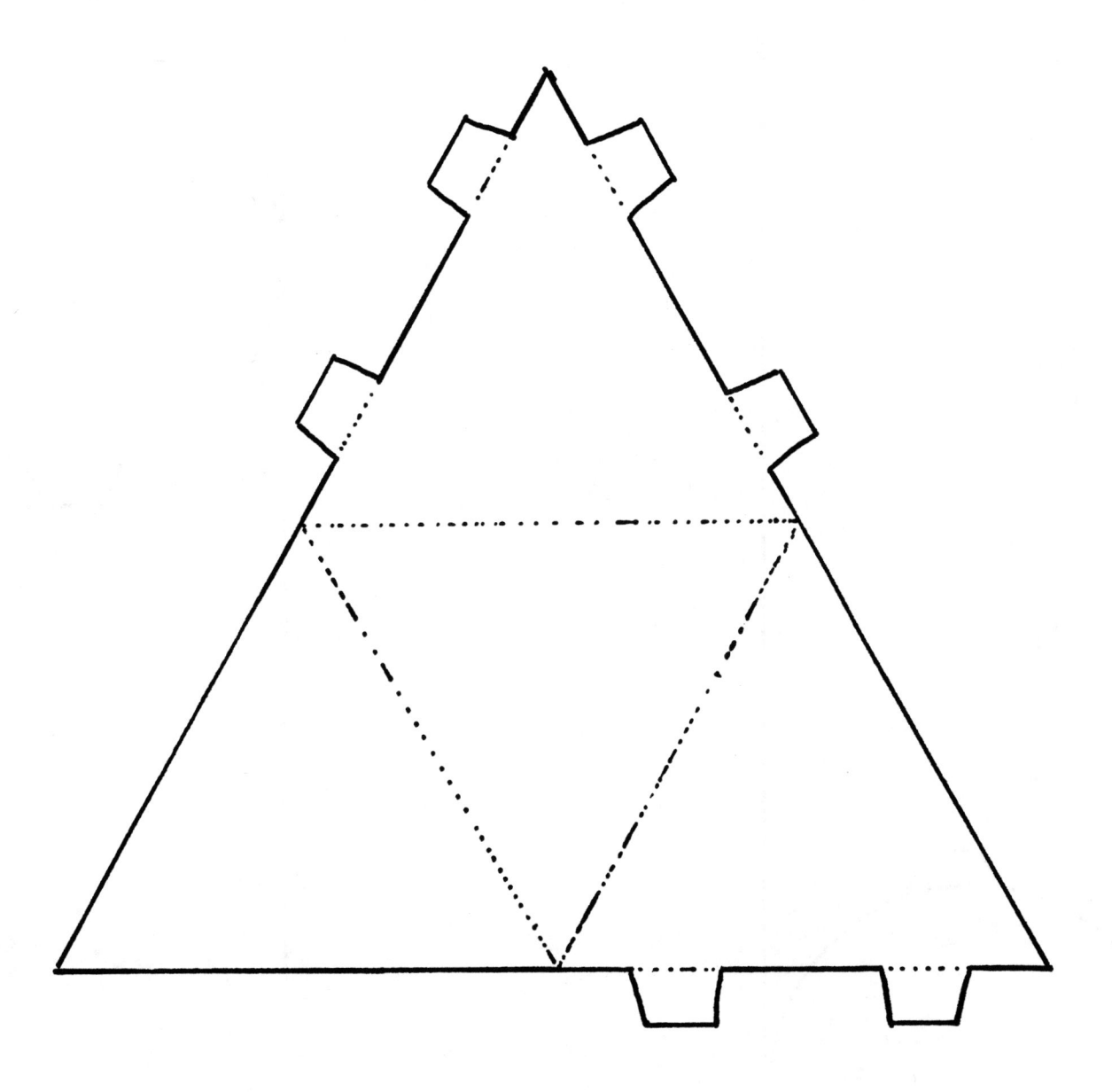

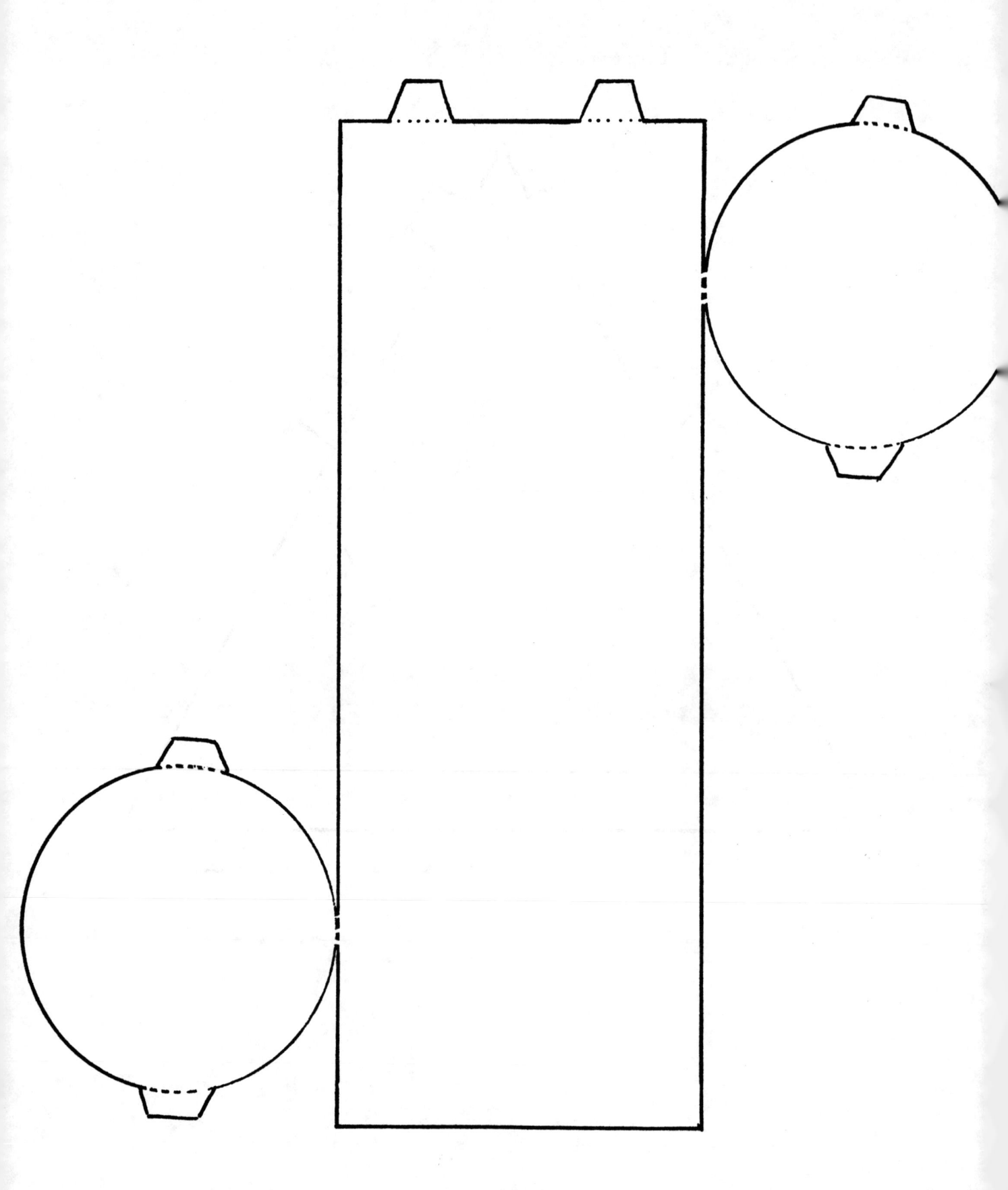

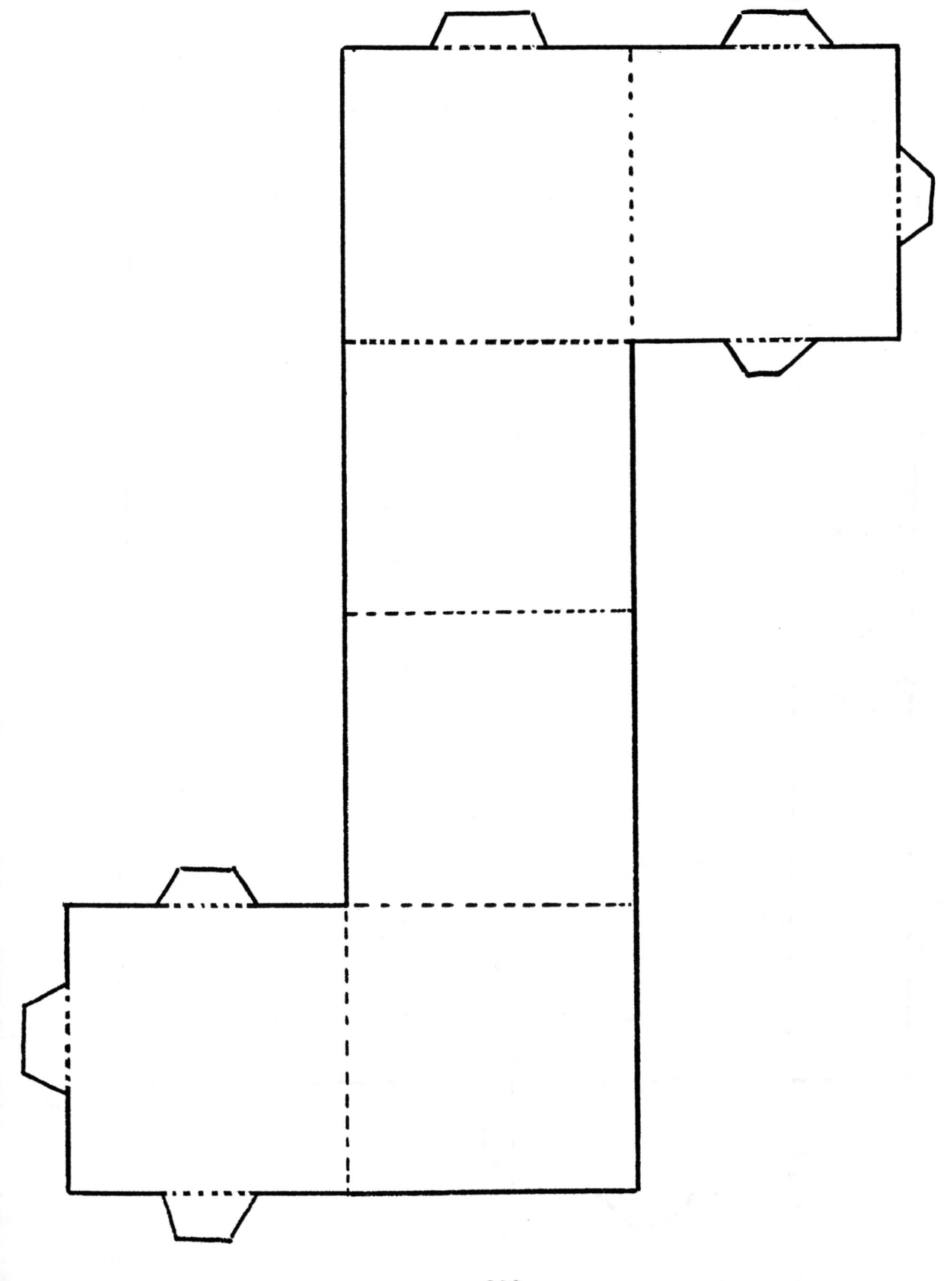

※

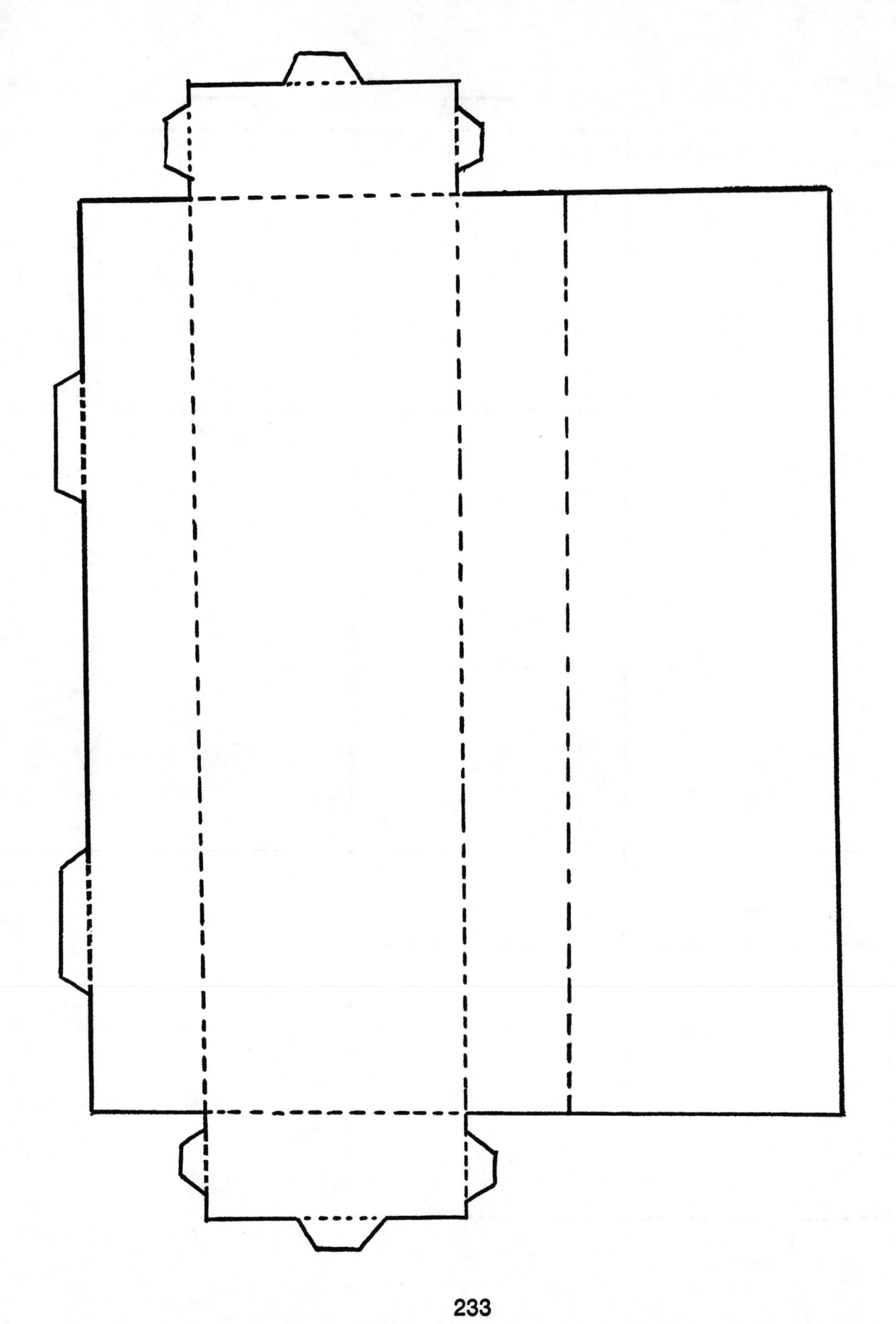

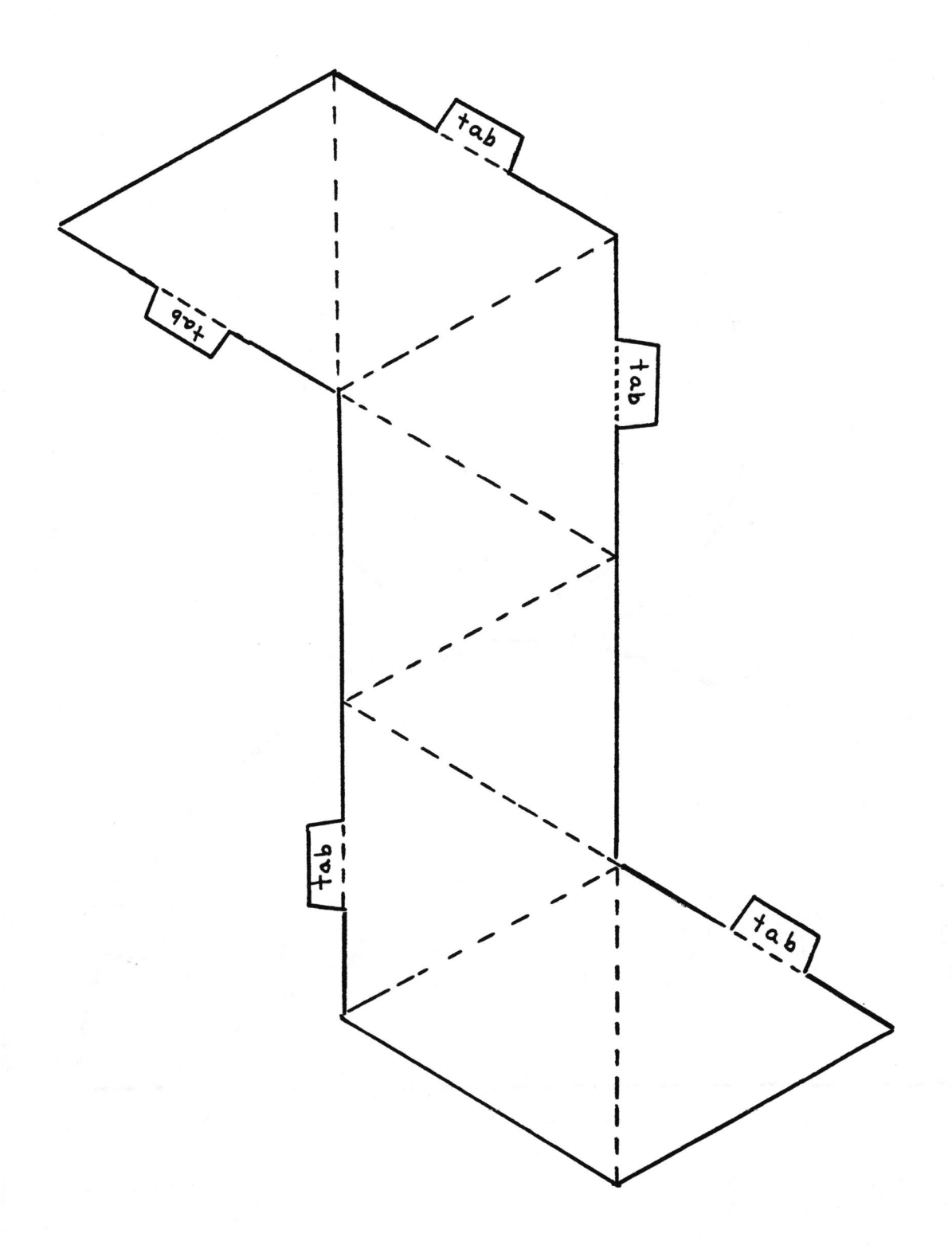

# SYMME-TREE

Skill: *Identifying symmetrical figures*

1. "Plant" a large (dead, please) branch in a can of plaster or gravel.

2. Draw several figures or pairs of figures on leaf-shaped pieces of construction paper. Make many of the figures or pairs symmetrical.

3. Fasten a string or pipe cleaner hook to each leaf. Keep the leaves in a small nest or basket in the "tree."

4. A student examines each figure to decide if it is symmetrical, and if so, hangs it on the tree.

5. Ask each student to add one original leaf to the tree.

## DOUBLE UP

---
Skill: *Drawing symmetrical figures*
---

1. Create a symmetry center where students may examine figures and draw reflections of what they see.

2. Provide several small mirrors and an assortment of figures. Label each.

3. The student takes down a figure, traces it on his paper, places a mirror on the edge of the shape, and draws its reflection.

## THIS IS SYMME-RAMA!

**Skill:** *Creating symmetrical designs*

Show students how to create designs, using the principles of symmetry. Here are some ideas!

Fold a 9 x 12-inch piece of colored construction lengthwise.

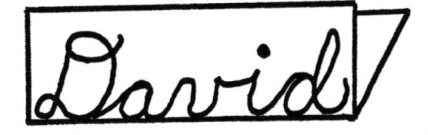

With chalk, write your name so that the letters touch the fold. If one letter has a tail, leave it off. Connect all letters.

Cut around the name, leaving the two sides joined at several points along the fold. (See illustration below.)

Open the "name insect" and glue on a contrasting piece of paper.

Fold a piece of paper. Open it. Dribble paint heavily on one side of the fold. Close, and rub outward from the fold. Open immediately. When almost dry, repeat with another color.

Cut four 6 x 18-inch pieces of construction paper (two white, two black). Fasten together one piece of each color by gluing along the long edges.

Cut identical designs from the other two pieces by holding them together while cutting. Begin all cuts at one of the long edges. When designs are cut, paste the white shapes on the black half of your two-toned sheet, and the black designs on the white half, making the pieces meet at the center line.

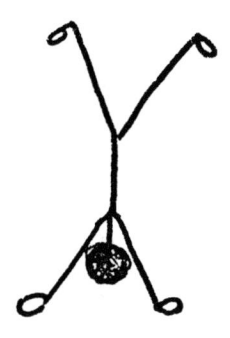

SLIDE, FLIP, OR TURN?

---
Skill: *Recognizing and drawing a plane figure after a slide, a turn, or a flip*

---

1. Make three charts, one to explain each of the terms *slide*, *turn*, and *flip*. (See the following page for samples.)

2. Attach an envelope to each chart in which you have placed patterns of the figures shown. This enables the student to move the figure and examine its appearance after the move.

3. On a fourth chart, draw several figures. Repeat many of the same figures, flipping or moving them to different positions. Label each figure.

4. Students study and experiment with the three charts. Then have them use Chart 4 to do such tasks as:

   a. *Find the flip of Figure A.*
   b. *Draw G after a vertical slide.*
   c. *Is B the same as M after 3/4 turn?*
   d. *Which figure is a flip of D?*

## What is a SLIDE?

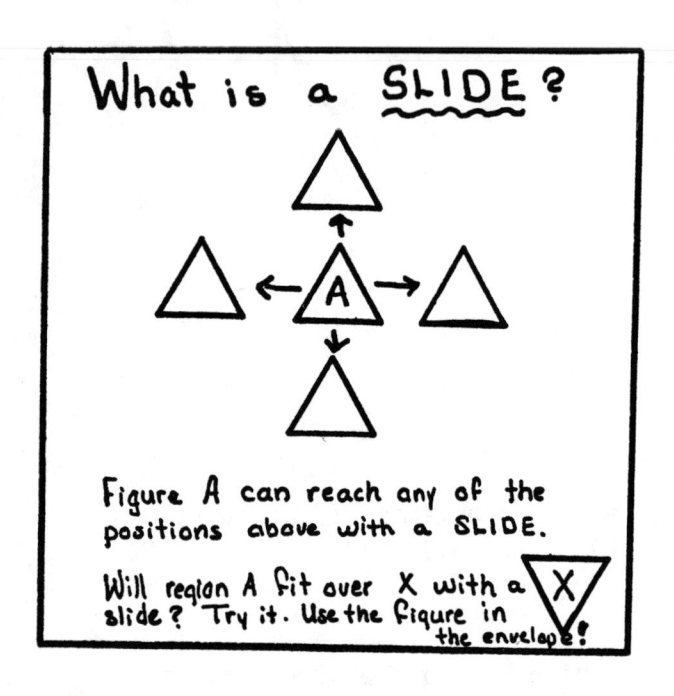

Figure A can reach any of the positions above with a SLIDE.

Will region A fit over X with a slide? Try it. Use the figure in the envelope!

## What is a TURN?

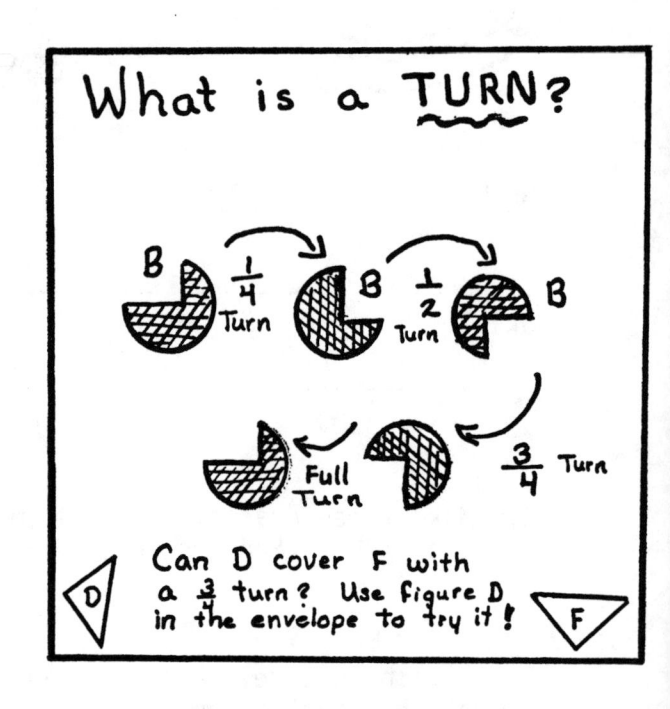

Can D cover F with a 3/4 turn? Use figure D in the envelope to try it!

## What is a FLIP?

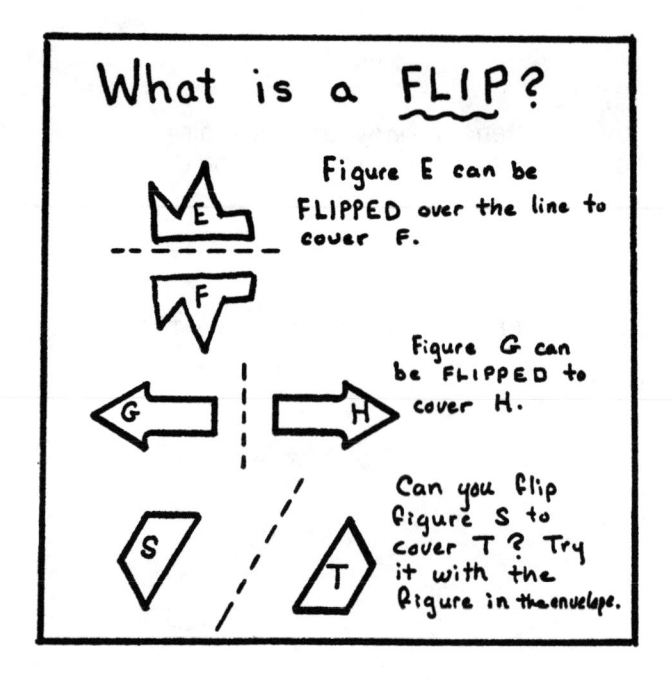

Figure E can be FLIPPED over the line to cover F.

Figure G can be FLIPPED to cover H.

Can you flip figure S to cover T? Try it with the figure in the envelope.

## SLIDE, FLIP, or TURN?

241

REMEMBER ME??

---
Skill: *Recognizing a figure in a different position*
---

1. Make a deck of 36 figure cards:

    a. Draw a figure on each of 16 cards.

    b. On the next 16 cards, draw the same figures, but place them in different positions (after sliding, turning, or flipping).

    c. The last four cards should be figures for which there are no matching cards.

2. Place the card game in a box. Include the following directions for players:

    a. This game accommodates two to four players.

    b. Deal six cards to each player. Leave the remaining cards face down in the center.

    c. You will be trying to match as many pairs of figures as you can.

    d. Players take turns, beginning with the dealer, moving clockwise.

    e. At each turn, a player may draw one card from the deck OR from the hand of the player to his left. As he matches a pair, he lays it aside.

    f. Play continues until all players have used all their cards except the unmatchables!

        SCORING:  5 points for each pair of figures
                  5 points to the first player to use all his cards
                  2 points for each unmatched card left at end of game

        THE FIRST PLAYER TO REACH 50 POINTS IS THE WINNER!

ELEPHANT, ƎLEPHANT

---
Skill: *Recognizing a figure in a different position*

---

1. Give students copies of the puzzle on the following page, or make a similar puzzle in a larger size for a bulletin board.

2. Supply students with a list of questions such as these:

   a. *Can B cover CC after one-half turn?* (No)

   b. *Which figure is a flip of A?* (X)

   c. *Can you find the slide of R?* (C)

   d. *What figure could cover H after a flip and a three-fourths turn?* (D)

   e. *Is E a slide of T?* (No)

   f. *Which figure could cover F after one-half turn?* (J)

   g. *Is there a figure that can cover G after a slide?* (No)

   h. *Which figure could cover N after one-fourth turn?* (K)

   i. *Could I cover L after a slide?* (Yes)

   j. *Which figure could cover M after a flip?* (P)

   k. *Is Q a flip of V?* (No)

   l. *Can you find a flip of W?* (No)

   m. *Is MM a flip of KK?* (No)

   n. *Find a slide of ZZ.* (TT)

*

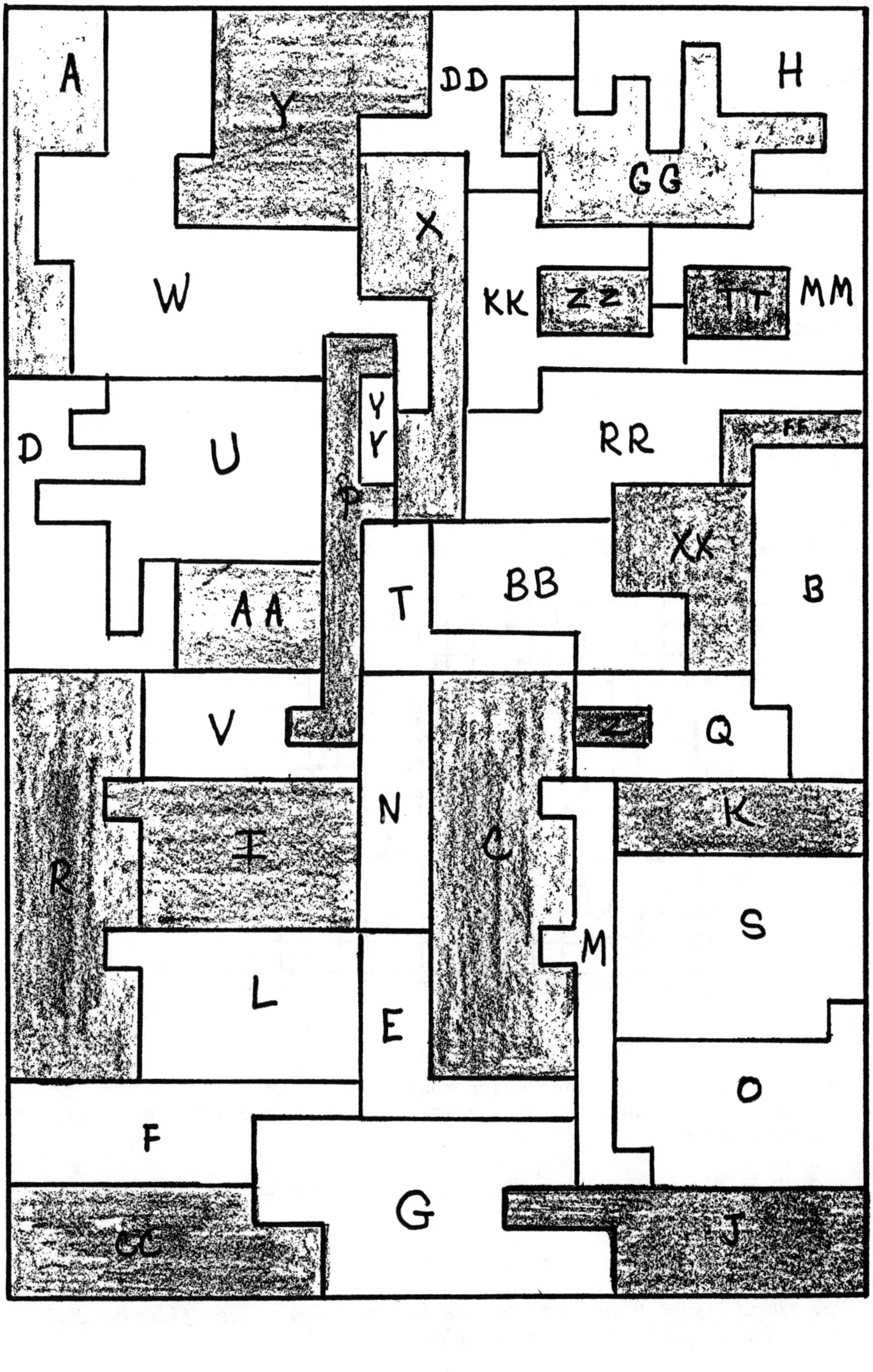
A
Y
DD
H
GG
X
W
KK
ZZ
TT
MM
YY
D
U
RR
FF
P
XX
T
BB
B
AA
V
Z
Q
N
K
R
I
C
S
M
L
E
O
F
G
CC
J

# GRIDOGRAPHY

---
Skill: *Moving figures on a grid*
---

1. Use lengths of yarn to form a coordinate grid on a bulletin board.

2. Number the axes with whole numbers.

3. Make several closed figures of various shapes. Label each figure with a letter.

4. Pin the figures at different locations on the grid.

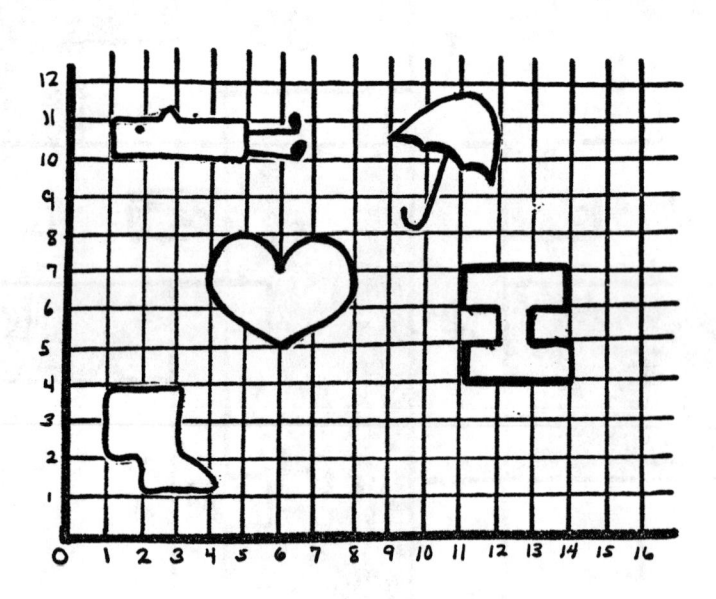

5. Post a list of activities which require students to change the locations of some figures. (See examples.) Provide graph paper, pencils, rulers, and crayons.

   a. *Draw Figure A on your grid. Draw it after it has been moved down 3.*

   b. *Draw Figure C on your grid as it would look after one-fourth turn.*

   c. *Draw Figure H on your grid as it would look after flipping and sliding four spaces to the right.*

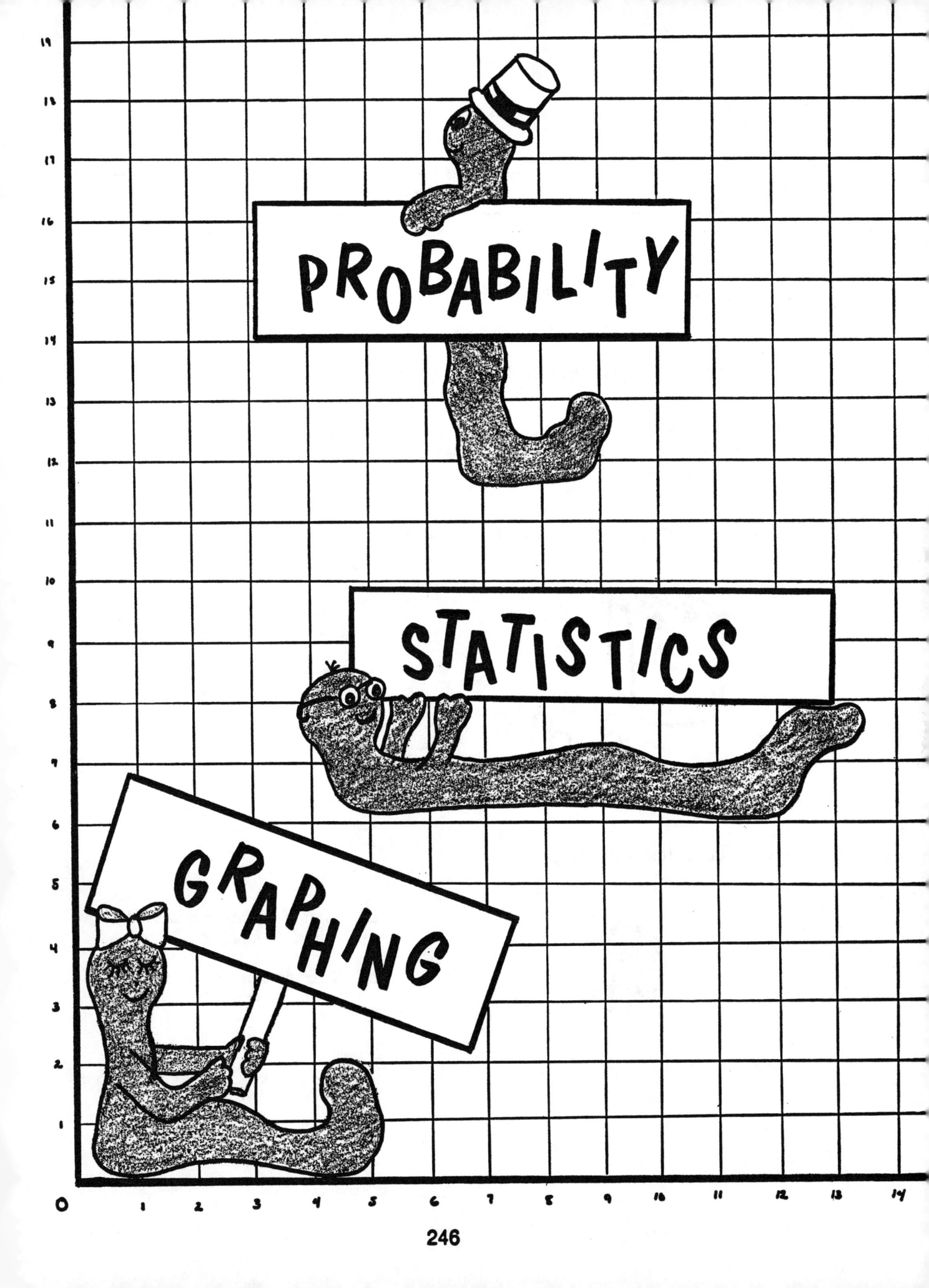

# PROBABILITY

# STATISTICS

# GRAPHING

CHANCE-IT!

---
Skill: *Using a ratio to express the probability of chance events*
---

1. Explain probability to the students:

    *the likelihood or chance that something will happen*

    Show them how to express probability with a ratio.

2. Prepare several Chance-It stations or boxes which students may visit to determine probability. At each station, put different materials and some suggestions for questions:

Box 1 contains a sack of colored marbles.

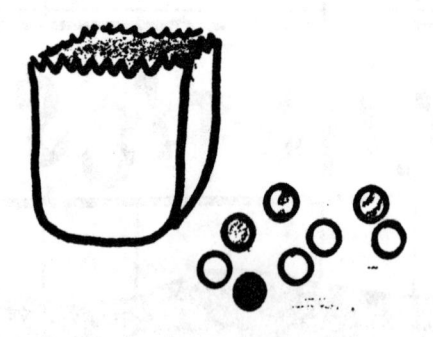

    a. What is the chance of drawing a blue marble?

    b. What is the chance of drawing a marble that is not red?

Box 2 contains a penny.

    a. What is the probability of tossing heads?

    b. In ten tosses, what is the probable number of tails?

Box 3 contains a numbered spinner.

a. What is the probability of spinning 4?

b. What is the chance of not spinning 3?

c. In three spins, what is the chance of spinning one 6?

Box 4 contains two pairs of shoes.

a. What is the chance of choosing a sneaker?

b. What is the chance of choosing a pair if you pick two shoes?

c. What is the probability of choosing an unmatched pair if you pick two?

Box 5 contains seven pencils (three red ones and four yellow).

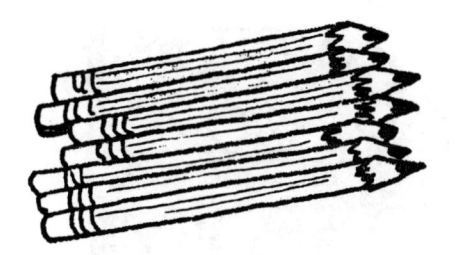

a. What is the chance of getting a yellow pencil?

b. What is the chance of not getting yellow?

3. Direct students to visit each Chance-It station and write a ratio in answer to each question.

   Variation: Students may revisit the stations later and conduct a number of trials, comparing the outcomes to their original probability ratios.

HOW DO THEY FALL?

---
Skill: *Predicting and recording outcomes in probability experiments*
---

1. Supply several pairs of dice. Label one die "A" and one "B."

2. Give each student a copy of the table on the following page.

3. Ask the students to list all the possible ways that a pair of dice might fall. Complete the first column of the table together as a group.

4. The students then determine the probability for each outcome in a series of fifty tosses.

5. Let students experiment on their own, tossing fifty times and recording their outcomes on their individual tables.

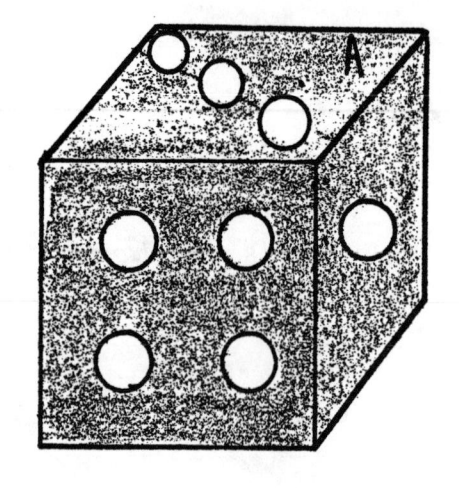

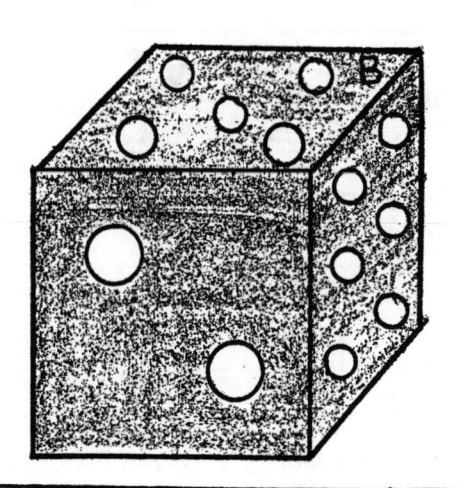

### How Do They Fall?

Sandi R.

| Combination | Probability Ratio | Outcomes | Ratio of Outcomes to Trials |
|---|---|---|---|
| (3, 5) | $\frac{1}{36}$ | III | $\frac{3}{50}$ |

# How Do They Fall?

Name

| Combination A : B | Probability Ratio | Outcomes | Ratio of Outcomes to Trials |
|---|---|---|---|
| | | | |

JACK-IN-THE-BOX?

Skill: *Using a ratio to express outcomes of probability experiments*

1. Cut a hole in the end of a shoebox. Make the opening large, so that a hand could reach into the box.

2. Inside the box, place photographs or names of five students (some boys, some girls).

3. Provide a sheet of directions and questions for students. Ask them to write probabilities for occurrences, as well as to express the outcomes of their trials.

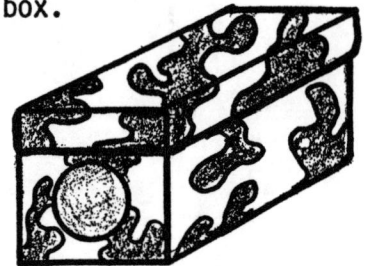

*What is the probability of choosing Randy?*
*What is the probability of choosing Cheri?*
*What is the probability of choosing a girl?*
*If you draw two photos, what are the possible combinations that might result?*
*What is the probability of choosing Jack and Doug?*
*List the probability for choosing each individual in 20 draws.*
*Now, draw, making sure that all the photos are in the box each time. Record your occurrences. Use a ratio to show the events. How do the ratios compare with your predictions?*

bright idea!

The seats in a theater or rows in a stadium are ready-made settings for beginning to work with coordinates. Draw a map of a theater or stadium. Have the students put names or people in the seats. Then, prepare questions for locating specific seats.

"Where is the lady in the red hat?"
"Give the coordinates of a blonde in the stadium."
"Where are the members of the Miller family seated?"

# US BEHIND BARS

**Skill:** *Making bar graphs*

1. Ask each student to select one "Statistics Assignment," which consists of gathering numerical data on a specific topic. Some topics are suggested below, but these may be varied for different grade levels. Encourage students to suggest their own ideas.

    *Heights of People in This Class*
    *How Fourth Graders Spend Their Allowances*
    *Favorite Records of This Class*
    *How Many Calories I've Eaten This Week*
    *How Much Time We Spend Watching TV*
    *Ages of the Teachers in Our School*

2. Provide some guidelines for collecting, organizing, and graphing data, reminding students to:

    *Keep a record of the information found.*
    *List the scores in order from highest to lowest.*
    *Decide on the interval for their axes before writing the numbers.*
    *Write numbers along left vertical edge and categories along bottom.*
    *Label the numbers and categories.*
    *Give a title to the graph.*
    *Make a key if necessary.*

3. Supply poster board, markers, crayons, yard or meter sticks, and rolls of wide ribbon.

4. When each student has collected his information, he pictures it on a graph, using ribbon to make horizontal or vertical bars.

5. Each student may formulate five questions which could be answered from his graph. The answers may be displayed with the graph.

6. When graphs are finished, students practice using bar graphs by answering the questions on others' graphs.

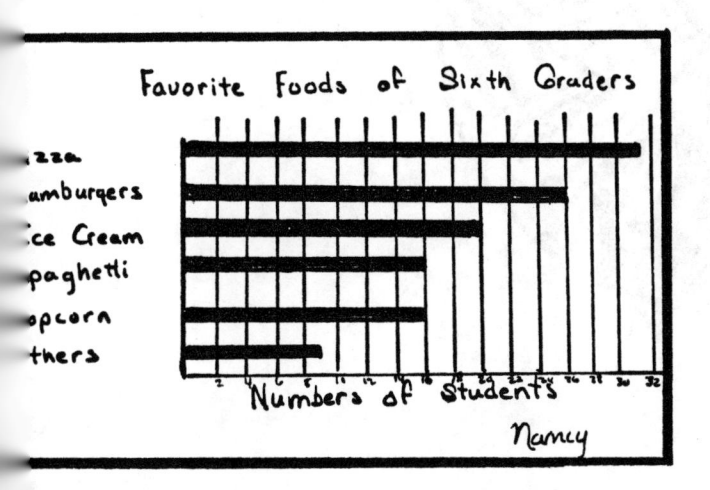

**Variation:** If you are working with students at a lower ability level, make the graph a group project. Each individual might be responsible for making one bar and writing one question.

## DAY IN THE ROUND

---
Skill: *Making circle graphs*
---

1. Ask students to keep a record of how they spend their time during one school day. (Older students may keep personal time calendars for a week.)

2. Collect cardboard "pizza" circles. Use pencil marks to divide them into six equal parts (if your school day is six hours long).

3. Give students paper plates on which to plan their circle graphs. This will enable them to experiment with proportions.

4. Use the heavier cardboard circles for the final graph. Students may color the different sections with crayon or paint, OR they may cut wedges from colored fabric. Each section should be labeled with a category and some indication of the time used. More advanced students may figure a ratio or percentage for each category.

## LINES TELL THE STORY

---
Skill: *Making and reading line graphs*

---

1. Cover a wall with a piece of mural paper (at least five yards long).

2. Have students draw lines to form a grid that is about 30 intervals high and 35 long. (These may vary according to numbers of students in your school.

3. Label the horizontal axis with the days of any particular month, and the vertical axis with numbers 0-30( or higher).

4. Choose a month for which the class will collect data on absences. Each day, send a student or a committee to find the number of absentees in each class.

5. A second committee may record the information on a table, and a third committee may transfer the data to the graph. Students may designate a different colored line for each class or each grade level,

    AND

    they might wish to tally and graph tardies also--perhaps by using dotted lines.

6. Remind students to include a title, key, and labels for the graph.

7. At the end of the month, have students formulate questions concerning the graphed data and display these along with the graph in the hallway. Your class and other may find the graph a useful part of math instruction!

# EGG HUNT

---
**Skill:** *Locating objects in the first quadrant of a coordinate grid*

---

1. To introduce coordinate graphing, prepare a simple grid on an overhead transparency. Draw colored eggs at several locations.

2. Cut a bunny out of cardboard (about six inches high). Fasten him to a stick or ruler.

3. Tell the students that the bunny is hiding eggs. Explain that as he hides the eggs, he always starts at zero and that he hops over (right) and then up.

4. Let the students take turns going to the screen with the bunny and giving a pair of numbers to locate the different eggs.

5. Place a new (empty) grid on the projector. Have students help to number the axes. Then, let them draw eggs at various locations as you say the coordinates.

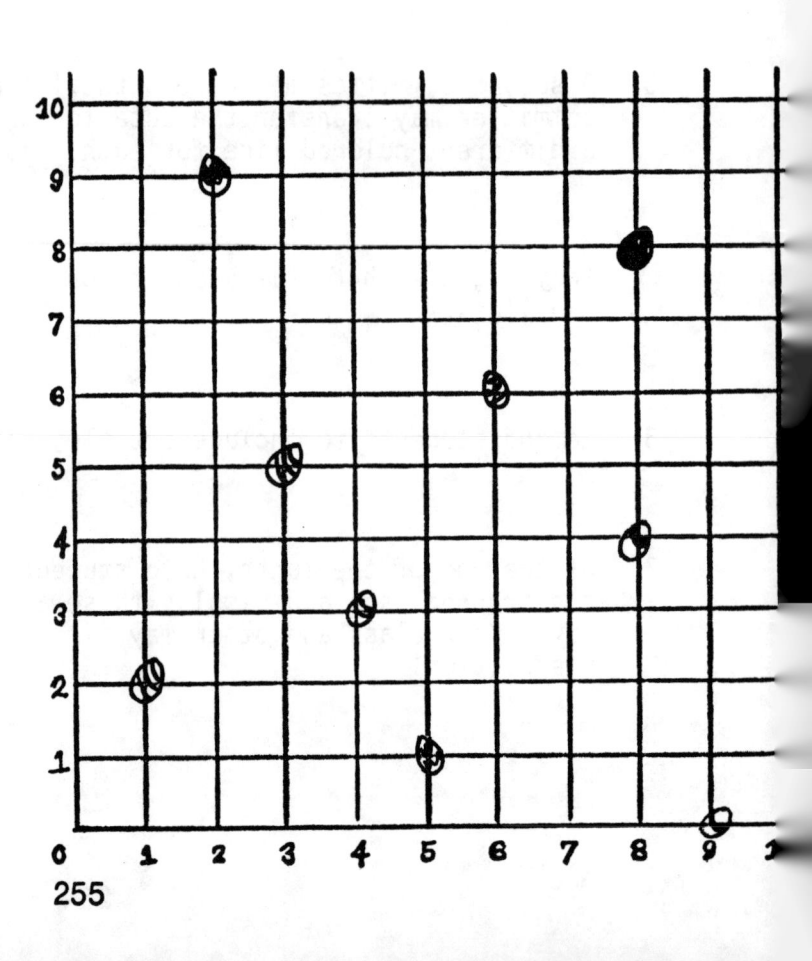

TIC-TAC-GRAPH

---
Skill: *Graphing an ordered pair of whole numbers*
---

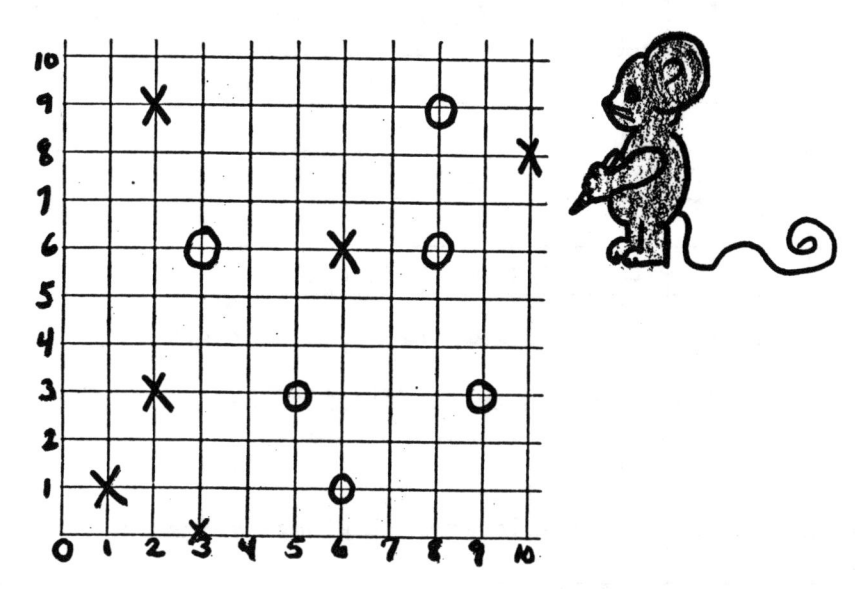

Variation: To make the game (below) easier, require a total of six marks in a row to win.

1. Use a piece of cardboard about one foot square to make a grid. Place the lines at one-inch intervals. Draw axes on the left and bottom edges, numbering them with whole numbers 0-10. Cover this with clear adhesive paper for durability.

2. Also make: an X and an 0 from bright fabric

    AND

    a set of 121 cards, each containing a pair of coordinates for graph.

3. Tic-Tac-Graph is a game for two players. One player pins the X on himself; the other wears the 0.

4. To play the game, players shuffle the cards and begin. At each turn, a player draws a card and places his mark at that location.

5. The first player to complete a horizontal, vertical, or diagonal row is the winner.

## A HAPPY FACE AT 20, 20!

---
**Skill:** *Locating objects in the first quadrant of a coordinate grid*
---

1. Transform your room into a grid! This will be very easy if your floor is tiled!

    *Use colored tape to mark the axes.

    *Measure to determine regular intervals.

    *Tape numbers to the floor or walls.

2. Let students locate all kinds of items:

        their desks            the reading rug
        friends' desks         the globe
        Ruth's sneaker       Tom's coat hook
        the doorknob         the record player

3. Assign some locating tasks:

    *Let's keep the wastebasket at (7, 12).*

    *Where shall we store the hockey sticks?*

    *The table at (10, 4) needs cleaning.*

    *A treat is waiting for you at (27, 14).*

    *Draw a happy face on the floor at (20, 20)!*

                              ☺

4. Try adding a second, third, and fourth quadrant to give practice with negative integers.

# GRID PICNIC

**Skill:** *Locating objects in four quadrants of a coordinate grid*

1. Find a checkered tablecloth (preferably plastic) or draw a checkerboard pattern on an old tablecloth or shower curtain.

2. Turn the tablecloth into a coordinate grid by drawing axes in the center. Number the areas with positive and negative integers.

3. Provide several figures and pictures that students may pin or tape to the correct location, thus creating a picnic lunch scene.

4. Tape record directions for placing the pictures:

    *Pin the apple at (-4, -2).*
    *The bread belongs at (5, 0).*
    *Put a plate at (-9, -10).*

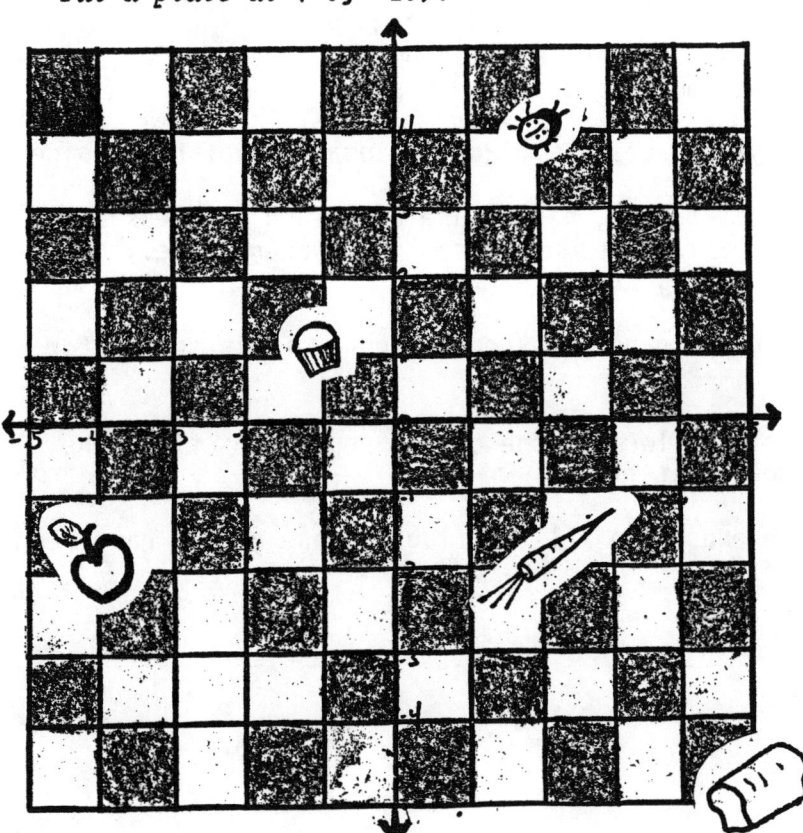

5. Supply a picture of the grid completed so that a student may check his work at a glance.

258

# CAPTURE THE FLAG

---
**Skill:** *Locating objects in four quadrants of a coordinate grid*

---

1. Obtain two pegboard squares (about a foot square) or make cardboard pegboards by poking holes at half-inch intervals.

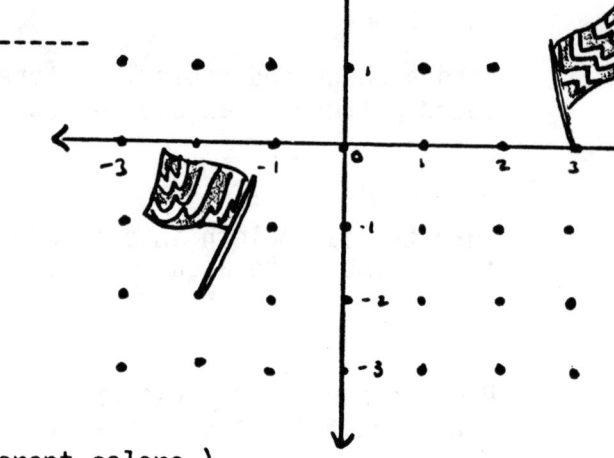

2. Make twenty flags by attaching fabric or paper squares to toothpicks. (Make ten flags of two different colors.)

3. Draw axes on the squares and number them with positive and negative integers.

4. To play the game, players need pegboards, graph paper, pencils, and a screen or notebook to separate players.

    a. Each player places his flags in locations of his choice.

    b. As the game begins, he may use each turn to guess a location of a flag belonging to his opponent, i.e., *"Do you have a flag at ( 7, -2)?"*

    c. He uses graph paper to keep a record of his search for the flags.

    d. When a player locates one of his opponent's flags, he "captures" that flag.

    e. The game ends when all the flags of one player have been captured.

GREAT GRID! I'M SUPERKID!

---
Skill: *Graphing ordered pairs of integers on a coordinate grid*
---

1. Supply students with graph paper and a set of coordinates which, when connected, reveal a message.

2. Instruct students to plot the points for each sub-set and connect those with a line before graphing the next group.
   For example:

   Graph (-11, 7), (-10, 5). Connect these points in order.
   Graph ( -9, 7), (-10, 5) (-3). Connect these points in order.
   ( This forms the first letter in the grid below.)

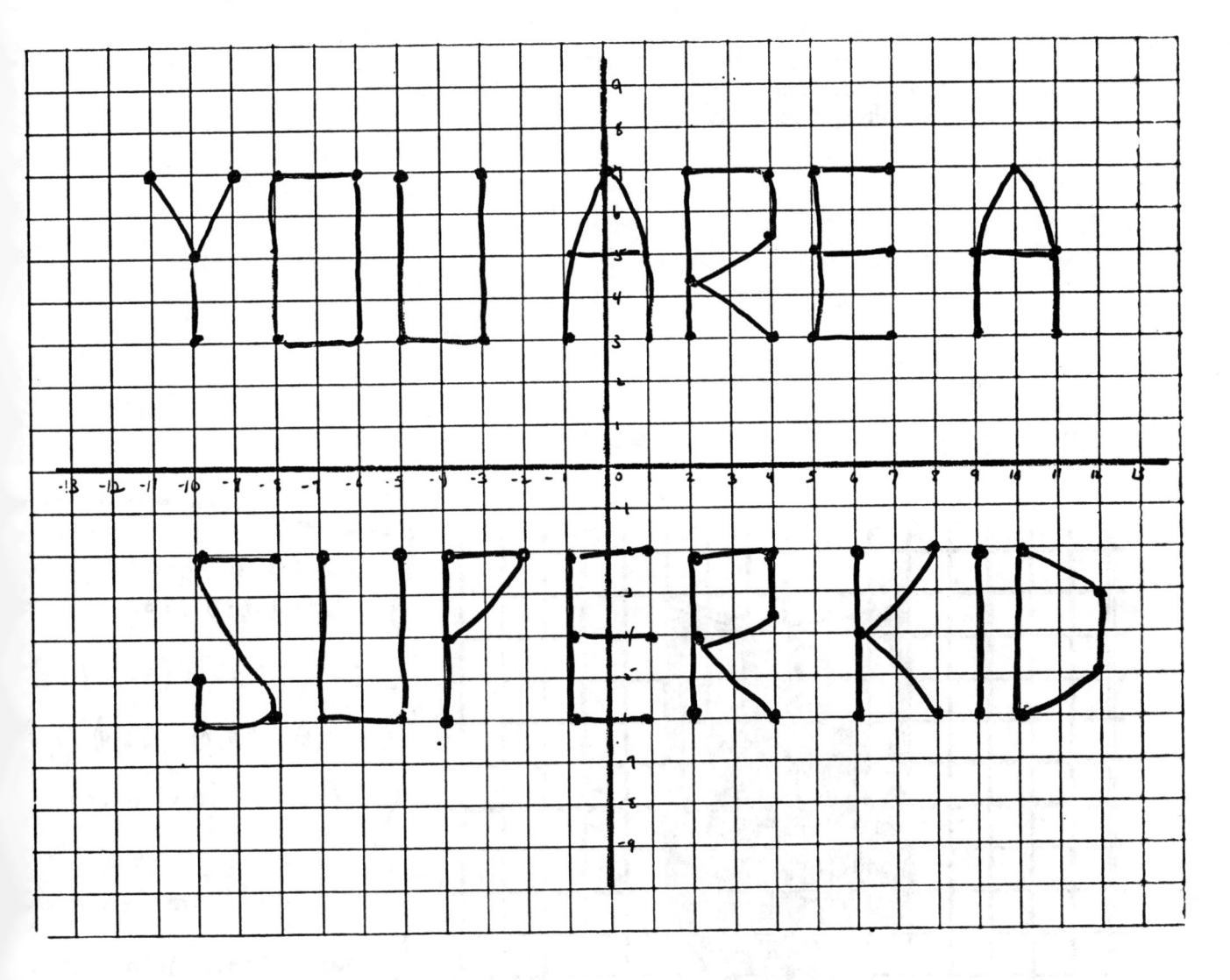

3. Ask each student to write coordinates for points which would spell his name. Then, when the papers are collected, you can graph the points to discover the name!

260

# GRAPH GALLERY

---
Skill: *Graphing ordered pairs of integers on a coordinate grid*
---

Graphing is such fun when the finished result is a picture! Here are the coordinates for a few.

Suggestions:

*To simplify the task, use only whole numbers.*

*Encourage students to make their own pictures and write down the coordinates for others to try. You could end up with a whole file of fun things to graph!*

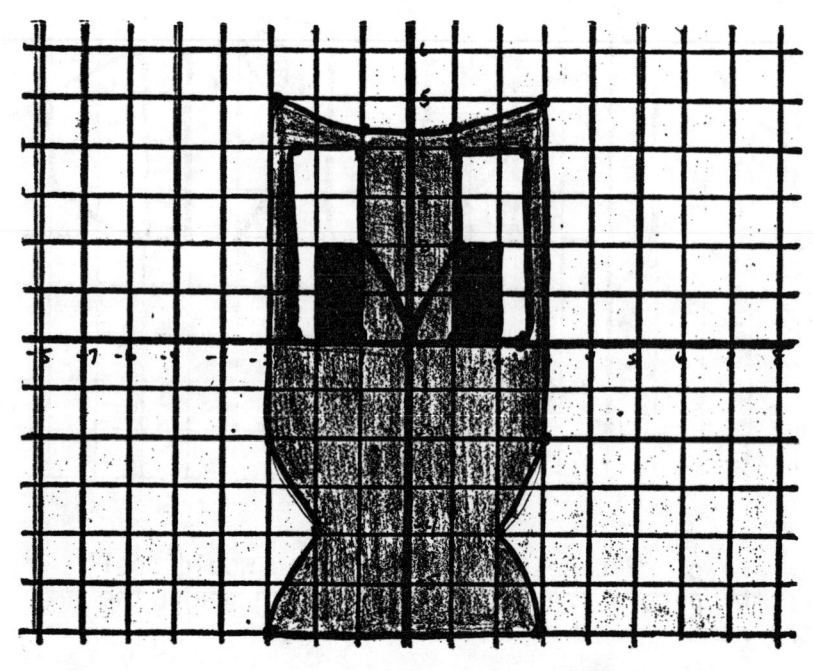

A (-3,5) (-3,-2) (-2,-4) (-3,-6) ( 
( 2,-4) ( 3,-2) (3,5)

B (-1,0) (-2½,0) (-2½,4) (-1,4)

C (1,0) (2½,0) (2½,4) (1,4)

D (-1,2) (0,0) (1,2)

Plot the points in each row, then join those with a line before graphing the next gro[up]

261

Connect these points with a line:
(3,4) (5,4) (5,3) (6,3) (6,4) (10,4)
(10,3) (11,3) (11,4) (14,4) (14,6)
(12,6) (12,8) (12,9) (3,9) (3,4)

Connect these:
(4,6) (6,6) (6,8) (4,8) (4,6)

Connect these:
(7,6) (9,6) (9,8) (7,8) (7,6)

Connect these:
(12,6) (10,6) (10,8) (12,8)

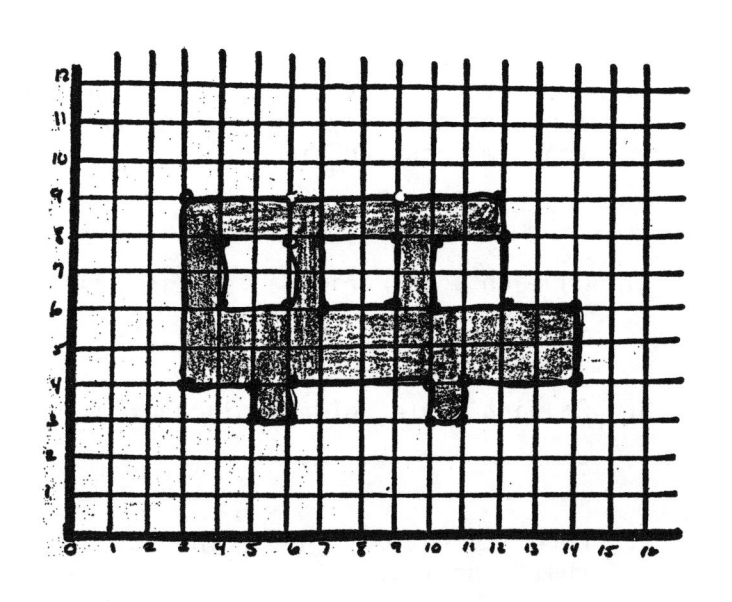

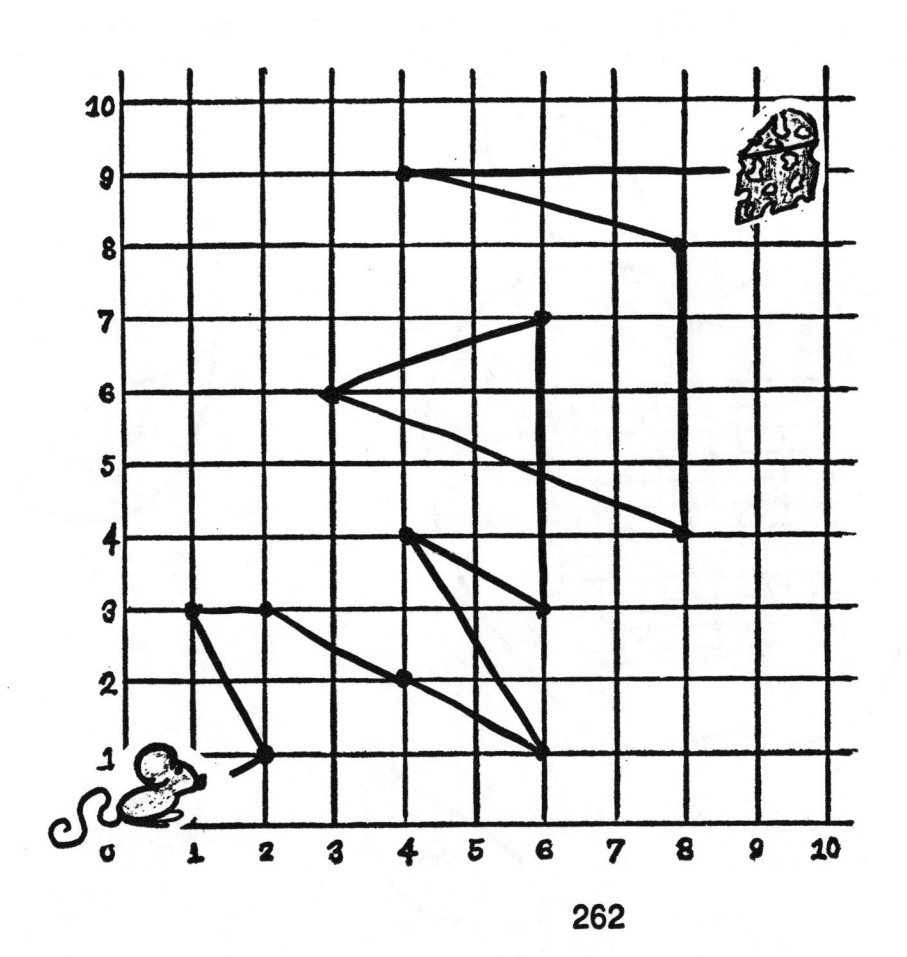

Help the mouse find the cheese!

Connect these points with a line!
(2,1) (1,3) (2,3) (4,2)
(6,1) (4,4) (6,3)
(6,7) (3,6) (8,4)
(8,8) (4,9) (9,9)

Be sure to connect the points in the order they're listed!

## GRAPH-APPLES

---
Skill: *Graphing functions*

---

1. Cut several leaves from green construction paper (at least enough for each student). Make, or have students make, large apples from red construction paper.

2. On each leaf, write a function rule.

3. On each apple, glue a piece of graph paper.

4. Each student may choose one function (leaf) and make a graph of that function.

5. Display the apples on a bulletin board. Pin the leaves in a cluster on the same board. Let students match the leaves with the corresponding apples.

(0, 1)
(1, 3)
(2, 5)
(3, 7)
(-1, -1)
(-2, -3)
(-3, -5)

(n x 2) + 1

n + 3

n - 4

# LINE DESIGNS

---
Skill: *Making line designs on a coordinate grid*
---

1. Collect a supply of:

    colored string          pencils
    needles                 rulers
    small, thin nails       hammers
    pieces of cardboard     graph paper
    old picture frames      white paper

2. Show students how a series of straight lines can look like a curve. Draw straight lines, connecting points on an X and Y axis. See the example below.

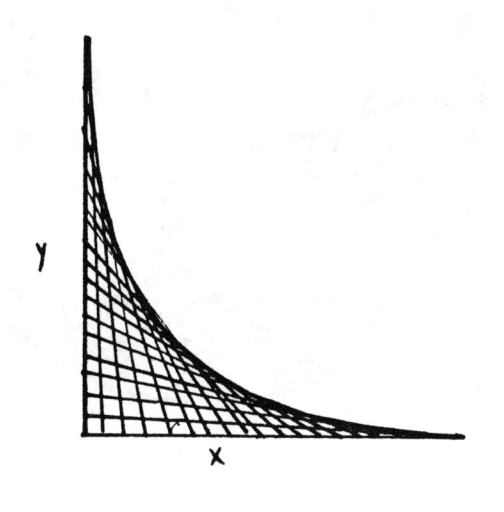

3. Let students use graph paper, rulers and crayons to experiment. The design will vary when:

    the shape of the grid changes.
    the size of the interval changes.

4. Students may create designs on paper with pencils and pens, or they may sew their designs on cardboard with colored string. Some may wish to try stringing the lines between nails hammered around the edges of old picture frames.

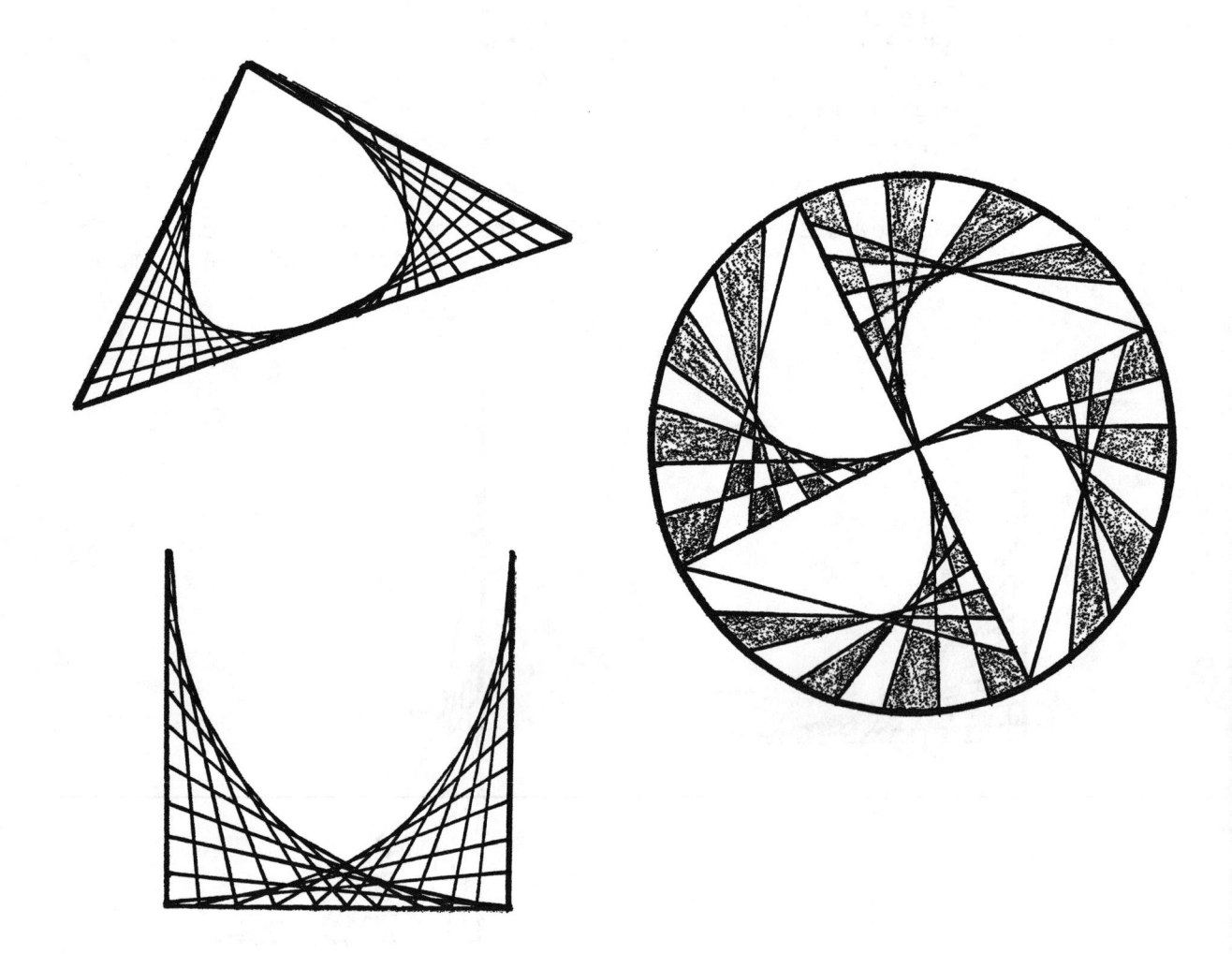

SUGGESTION!! You might wish to become the owner of a helpful little booklet called *Line Designs* by Dale G. Seymour and Joyce Snider. It is published by Creative Publications, Box 10328, Palo Alto, California 94303, and provides all kinds of exciting ideas for making designs with lines.

# APPENDIX

I. Skills That Make a Math Whiz

II. Terrific Tables of Measures, Formulas and Symbols

III. Treasures, Tools and Tidbits for Math Lovers

IV. Hung Up on a Math Word?
(A Glossary--Every Word You Ever Wanted to Know About Math!)

# skills that make a math whiz —

I. <u>Numeration</u>

Recognize that numerals name numbers
Associate word names with their corresponding numerals
Read and write one-digit numerals
Associate numerals with intervals along a number line
Read and write two-digit numerals
Identify place value for two-digit numerals
Identify numeration for one-half, one-quarter, and one-third
Express two-digit numerals in expanded notation
Rename tens as ones and ones as tens
Read and write three-digit numerals
Identify place value for three-digit numerals
Express three-digit numerals in expanded notation
Rename hundreds as tens and tens as hundreds
Identify values of a penny, nickel, dime, quarter, and dollar
Read and write numerals in a non-decimal base
Recognize the sequence of whole numbers
Read and write four-, five-, and six-digit numerals
Identify place value for four-, five-, and six-digit numerals
Express four-, five-, and six-digit numerals in expanded notation
Read and express numbers using an abacus
Read and write Roman numerals
Devise an original numeration system
Read and write numerals through the millions
Identify place value for numerals through nine digits
Express seven-, eight-, and nine-digit numerals in expanded notation
Recognize period grouping for nine-digit numerals
Rename thousands as hundreds
Distinguish between numerals having the same digits in different positions
Read and write numerals larger than one million
Read and write numerals for fractional numbers
Rename whole numbers as fractions and fractions as whole numbers
Read and write mixed numerals
Read and write decimal numerals
Express decimals in expanded numeration
Read and write mixed decimal numerals
Read and write positive and negative integers
Read and write exponential numerals
Read and write percentages

267

II. <u>Number Theory</u>

Identify even and odd numbers
Give the factors of whole numbers
Identify common factors of two whole numbers
Identify the greatest common factor of two whole numbers
Give the multiples of numbers
Identify common multiples of two whole numbers
Identify the least common multiple of two whole numbers
Identify prime numbers
Identify composite numbers
Determine if whole numbers are divisible by 2,3,4,5,6,9, or 10

III. <u>Number Concepts</u>

Order the numerals 1-10
Recognize the ordinal positions first through tenth
Recognize and use the concepts more, less; greater, smaller; few, many
Recognize a whole as greater than any of its parts
Understand the value and properties of zero
Compare and equalize the values of whole numbers
Recognize that there are many names for one number
Understand and use the concept of equal and the symbol =
Understand and use the concept of inequality and the symbols <, >, ≠
Read and write sentences which name a whole number
Read and write sentences using <, >
Complete unfinished sentences using <, >
Complete sequences of whole numbers
Order fractional numbers
Compare the values of fractional numbers
Complete sequences of fractions
Complete sequences of decimals
Order positive and negative integers
Complete sequences of integers
Round whole numbers
Round fractions
Round decimals
Find the value of numbers with exponents
Increase and decrease numbers by given units
Use words and symbols to represent numbers

IV. <u>Sets</u>

Classify objects or items into sets
Associate numerals and names with sets of corresponding numbers
Write numerals to correspond with sets
Associate the symbol 0 with an empty set
Seperate members of sets into subsets
Match members of two sets in one-to-one correspondence
Group items into sets of 2, 5, and 10
Given two sets, identify which is greater and which is less
Identify equivalent sets

Identify non-equivalent sets
List members of sets, using braces
Use the terms equivalent and non-equivalent and the symbols
Arrange sets in order of size
Find the union of two or more sets
Find the intersection of two or more sets
Use the symbols ∪ and ∩
Identify the set of whole numbers
Identify the set of integers
Identify the set of fractional numbers
Use fractions to name parts of sets
Form Cartesian sets
Use Venn diagrams to represent sets
Distinguish between finite and infinite sets
Identify solution and replacement sets


V.  Whole Numbers

   A. Addition and Subtraction

   Understand and use + and - symbols
   Discover and learn sums through 10
   Discover and learn differences through 10
   Recognize the inverse relationship between addition and subtraction
   Use terms: addend, sum, difference
   Discover and learn sums and differences through 20
   Recognize zero as the identity element for addition
   Use the commutative property for addition
   Find the missing addend in addition sentences
   Use a number line to find sums and differences
   Use the associative property for addition
   Add and subtract problems written vertically
   Find sums and differences through 100
   Estimate sums and differences
   Rename tens as ones
   Add and subtract two-digit numbers with renaming
   Rename hundreds as tens
   Add and subtract three-digit numbers with renaming
   Subtract three-digit numbers with zeros
   Add long columns of numbers
   Rename thousands as hundreds, ten thousands as thousands, and
       hundred thousands as ten thousands
   Add and subtract with more than four digits
   Write related addition and subtraction sentences
   Add and subtract in non-decimal systems

   B. Multiplication and Division

   Group objects into equivalent sets
   Separate sets into equivalent subsets
   See multiplication as the joining of equivalent sets
   See multiplication as repeated addition
   See division as separation of sets into equivalent subsets

See division as repeated subtraction
Recognize inverse relationship of multiplication and division
Write multiplication and division sentences for numbers in sets
Discover and learn multiplication and division facts using
    factors through 5
Use terms: factor, product
Use a number line to solve multiplication and division problems
Recognize one as the identity element in multiplication
Supply missing factors in multiplication and division problems
Recognize and use the commutative property for multiplication
Discover the role of zero in multiplication and division
Discover and learn multiplication and division facts through 100
Recognize and use the associative property for multiplication
Write and solve word problems involving multiplication and di-
    vision facts through 100
Multiply by tens and multiples of ten
Divide by ten and multiples of ten
Estimate products and quotients
Recognize and use the distributive property for multiplication
    over addition
Multiply two-digit numbers by a one-digit multiplier
Divide two-digit numbers by a one-digit divisor
Use terms: quotient, divisor, remainder
Multiply three or more digit numbers by a one-digit multiplier
Divide three or more digit numbers by a one-digit divisor
Multiply by two-digit multipliers
Multiply by three-digit multipliers
Multiply by multipliers larger than three digits
Check division problems using multiplication
Find averages
Check multiplication by casting out nines
Divide by 100 and multiples of 100
Divide by two-digit divisors
Divide by three-digit divisors
Divide by divisors larger than three digits

VI. <u>Integers</u>

Identify positive and negative integers
Associate positive and negative integers with intervals along a
    number line
Recognize points above and below zero on a thermometer
Put positive and negative integers in order
Add integers
Subtract integers
Multiply integers
Divide integers

VII. **Fractional Numbers**

  A. **Fractions**

   Identify halves, fourths, thirds
   Recognize that two halves make a whole
   Use fractions to name parts of sets
   Identify the numerator and denominator of fractions
   Add like fractions
   Subtract like fractions
   Recognize equivalent fractions
   Use the term equivalent and the symbol
   Identify non-equivalent fractions
   Find the greatest common factor for the numerator and
      denominator of fractions
   Express fractions in lowest terms
   Identify improper fractions
   Identify mixed numerals
   Express improper fractions as mixed numerals
   Express mixed numerals as improper fractions
   Find the least common multiple for the denominators of
      two or more fractions
   Rename fractions as like fractions
   Add and subtract unlike fractions and mixed numerals
   Order fractions
   Multiply fractions
   Multiply mixed numerals
   Name the reciprocals of fractions
   Divide fractions, using the reciprocal method
   Identify complex fractions
   Divide fractions, using the complex fraction method
   Divide mixed numerals
   Use fractions to name ratios
   Use fractions to show probability
   Write and solve proportions
   Write ratios as percentages
   Write percentages as ratios
   Find the percentage of numbers

  B. **Decimals**

   Use dollar signs and decimal points to write amounts of money
   Add and subtract with money
   Multiply and divide with money
   Write decimals including tenths and hundredths
   Write decimals including thousandths
   Order decimals
   Add and subtract decimals
   Multiply decimals
   Divide decimals
   Express fractions as decimals
   Express decimals as fractions
   Express ratios as decimals
   Express decimals as percentages

VIII. <u>Problem Solving</u>

    Use manipulatives, pictures and dramatization to solve problems
    Complete number sentences corresponding to pictured problems
    Write and solve equations for pictured problems
    Answer questions and solve problems using information in pictures and simple graphs
    Complete number sentences corresponding to word problems
    Write and solve equations for word problems
    Formulate problems from information presented in pictures and word problems
    Estimate answers to word problems
    Determine what information is essential for solving a problem
    Identify information which is not essential for solving a problem
    Determine what operation is appropriate for solving a problem
    Choose a method for solving a problem
    Recognize more than one way to solve a given problem
    Solve word problems involving more than one step
    Combine two numbers in several ways
    Solve problems involving money
    Solve problems involving rate, time and distance
    Solve problems involving ratios
    Solve problems involving percentages
    Solve problems using information from tables and graphs
    Predict outcomes in problem situations
    Test various methods of solving a problem
    Solve equations with missing numbers
    Solve equations containing two variables

IX. <u>Measurement</u>

    Compare lengths and sizes
    Understand and use concepts: smaller, larger, more less
    Name days of the week, months of the year
    Recognize that there are 7 days in a week and 12 months in a year
    Tell time to the nearest hour
    Tell time to the nearest half hour
    Identify a dozen and a half-dozen
    Recognize that there are 24 hours in a day
    Recognize that there are 60 minutes in an hour and 60 seconds in a minute
    Tell time to the nearest minute
    Tell time to the nearest second
    Compare units in the English and Metric systems
    Measure length using inches, feet, yards
    Measure liquid capacity using liters
    Measure length using centimeters and meters
    Measure liquid capacity using cups, pints, qurts and gallons
    Measure temperatures using a thermometer
    Distinguish between a Farenheit and Celsius thermometer
    Identify freezing and boiling points on a Farenheit and Celsius thermometer
    Recognize that there are 52 weeks and 365 or 366 days in a year

Compare weights of objects
Measure weights using pounds or ounces
Use scales to find distances on maps
Draw maps or pictures to scale
Use a protractor to measure angles
Measure parts of circles in degrees
Use lines of latitude and longitude to measure distances on the
    earth's surface
Find the perimeters of regular polygons
Find the perimeters of irregular polygons
Measure the radius or diameter of circles
Find the circumference of circles
Find the area of rectangles
Find the area of triangles
Find the area of other polygons
Find the area of circles
Find the area of circles
Measure length using millimeters, decimeters, kilometers
Measure length using miles, rods
Measure liquid capacity using milliliters and kiloliters
Measure weight using milligrams and kilograms
Measure weight using tons
Find the volume of regular space figures
Find the surface area of polyhedra
Compare and convert measurements of temperature in Farenheit and
    Celsius degrees
Add, subtract, divide and multiply measurements
Estimate measurements
Measure with accuracy


X. Geometry

   A. Plane Geometry

      Identify circles, triangles, squares and rectangles
      Classify objects by shape
      Identify closed and open figures
      Recognize regions inside and outside closed figures
      Identify line segments
      Differentiate between a line and a line segment
      Measure and draw line segments of various lengths
      Name line segments AB
      Recognize and draw rays
      Name rays CD
      Construct congruent line segments
      Distinguish between simple closed figures and closed figures
          that are not simple
      Recognize parts of figures (halves, fourths, thirds)
      Identify intersection of line segments
      Identify parallel lines
      Identify perpendicular lines
      Understand and use the term plane
      Recognize and draw angles

Name angles
Recognize properties of circles, triangles and parallelograms
Identify congruent shapes
Identify similar shapes
Measure angles
Recognize acute, right and obtuse angles
Recognize congruent angles
Identify angles formed by the intersection of a line with two
    parallel lines
Identify parts of a circle: radius, diameter, center, arc,
    chord, tangent
Use a compass to draw circles
Find the circumference of circles
Find the area of circles
Identify quadrilaterals
Identify triangles
Identify other polygons
Find the perimeter of polygons
Find the area of polygons
Identify skew lines
Identify parallel planes
Identify intersecting planes
Construct parallel lines
Construct perpendicular lines
Construct a bisector for a line segment
Construct a bisector for an angle
Construct equilateral triangles
Construct congruent triangles

B. <u>Space Geometry</u>

Identify cubes
Identify other rectangular prisms
Identify triangular prisms
Identify pyramids
Identify cones
Identify cylinders
Identify spheres
Name and count faces, edges, vertices of space figures
Draw space figures
Construct space figures from paper
Find the surface area of prisms
Find the surface area of pyramids
Find the surface area of cones
Find the surface area of cylinders
Find the surface area of spheres
Find the volume of prisms
Find the volume of pyramids
Find the volume of cylinders
Find the volume of cones
Find the volume of spheres

### C. Transformational Geometry

Identify symmetrical figures
Draw symmetrical figures
Recognize the slide of figures
Draw the slide of figures
Recognize the turn of figures
Draw the turn of figures
Recognize the flip of figures
Draw the flip of figures
Recognize figures after several moves
Move figures on a grid

## XI. Probability, Statistics, and Graphing

Determine which of two events is more or less likely to happen
Determine the likelihood of chance events
Use a ratio to express probability
Record occurrence of events on a table
Record data on a graph
Find the mean in a set of data
Find the median in a set of data
Find the mode in a set of data
Read and make picture graphs
Read and make bar graphs
Read and make circle graphs
Read and make line graphs
Locate objects and points along a line numbered at regular intervals with whole numbers
Locate points along a line numbered at regular intervals with fractional numbers
Locate points along a line numbered at regular intervals with integers
Give the coordinates of pictures or points located in the first quadrant of a grid
Graph ordered pairs of whole numbers on a grid
Give the coordinates of points located in all four quadrants of a grid
Graph ordered pairs of integers on a grid
Graph solution sets to linear equations
Graph functions
Locate positions on the earth's grid

# terrific tables of measures, formulas, & symbols

## LENGTH

| | |
|---|---|
| 1 centimeter (cm) = 10 millimeters (mm) | 1 foot (ft.) = 12 inches (in.) |
| 1 decimeter (dm) = 10 centimeters | 1 yard (yd.) = 36 inches |
| 1 meter (m) = 10 decimeters | 1 yard = 3 feet |
| 1 meter = 100 centimeters | 1 rod (rd.) = 16½ feet |
| 1 decameter (dkm) = 10 meters | 1 rod = 5½ yards |
| 1 kilometer (km) = 1000 meters | 1 mile (mi.) = 5,280 feet |
| | 1 mile = 1,760 yards |

## WEIGHT

| | |
|---|---|
| 1 gram (gm) = 1000 milligrams (mg) | 1 pound (lb.) = 16 ounces (oz.) |
| 1 kilogram (kg) = 1000 grams | 1 ton = 2000 pounds |
| 1 metric ton = 1000 kilograms | |

## CAPACITY

| | |
|---|---|
| 1 liter (l) = 1000 milliliters (ml) | 1 pint (pt.) = 2 cups |
| 1 decaliter (dkl) = 10 liters | 1 quart (qt.) = 2 pints |
| 1 kiloliter (kl) = 1000 liters | 1 gallon (gal.) = 4 quarts |
| | 1 peck (pk.) = 8 quarts |
| | 1 bushel (bu.) = 4 pecks |

## TIME

| | |
|---|---|
| 1 minute (min.) = 60 seconds (sec.) | 1 year = 52 weeks |
| 1 hour (hr.) = 60 minutes | 1 year = 365 or 366 days |
| 1 day = 24 hours | 1 year = 12 months (mo.) |
| 1 week = 7 days | 1 decade = 10 years |
| | 1 century = 100 years |

FORMULAS

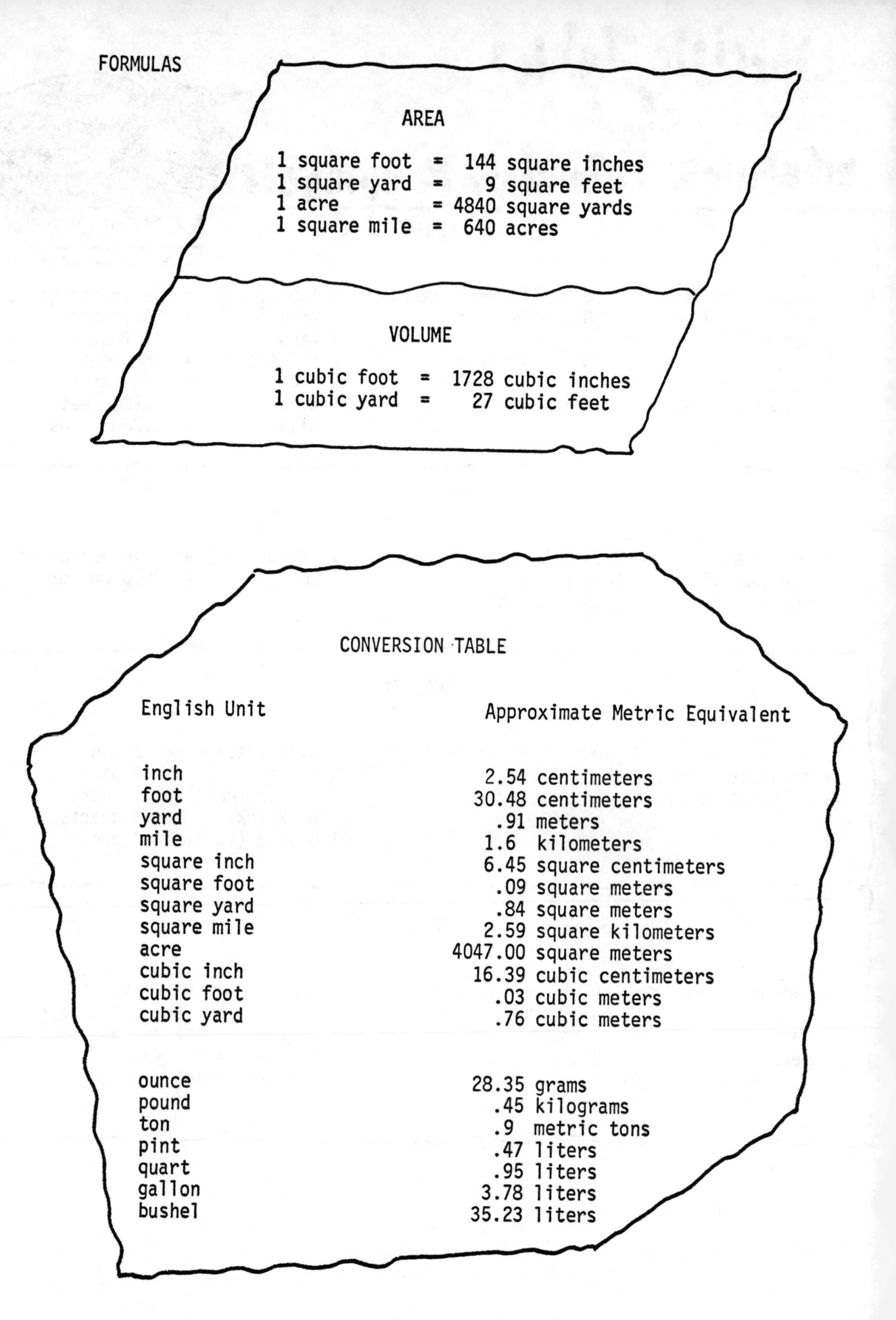

## AREA

1 square foot = 144 square inches
1 square yard = 9 square feet
1 acre = 4840 square yards
1 square mile = 640 acres

## VOLUME

1 cubic foot = 1728 cubic inches
1 cubic yard = 27 cubic feet

## CONVERSION TABLE

| English Unit | Approximate Metric Equivalent |
|---|---|
| inch | 2.54 centimeters |
| foot | 30.48 centimeters |
| yard | .91 meters |
| mile | 1.6 kilometers |
| square inch | 6.45 square centimeters |
| square foot | .09 square meters |
| square yard | .84 square meters |
| square mile | 2.59 square kilometers |
| acre | 4047.00 square meters |
| cubic inch | 16.39 cubic centimeters |
| cubic foot | .03 cubic meters |
| cubic yard | .76 cubic meters |
| ounce | 28.35 grams |
| pound | .45 kilograms |
| ton | .9 metric tons |
| pint | .47 liters |
| quart | .95 liters |
| gallon | 3.78 liters |
| bushel | 35.23 liters |

## AREA

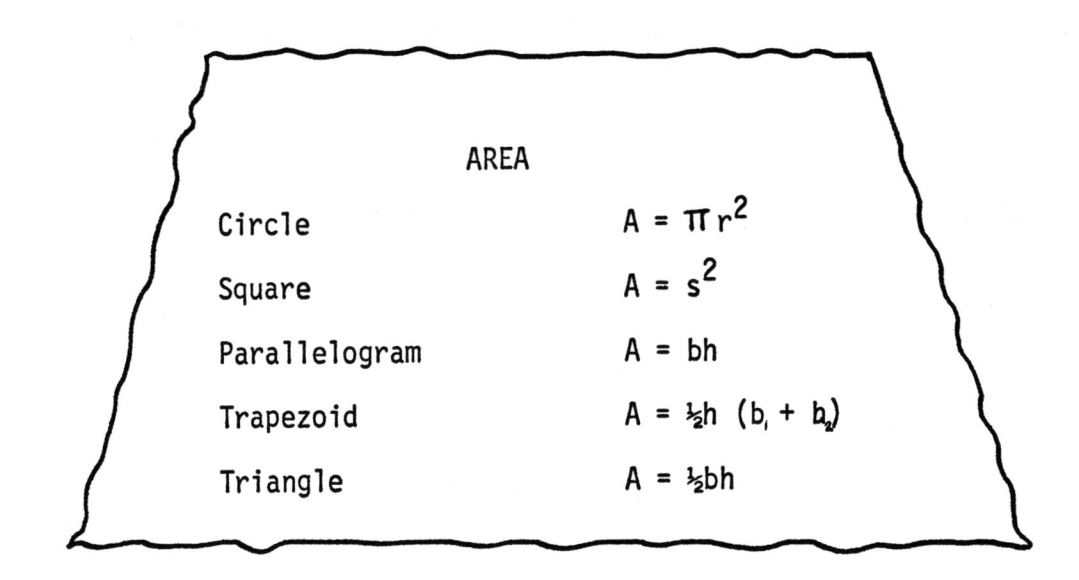

| | |
|---|---|
| Circle | $A = \pi r^2$ |
| Square | $A = s^2$ |
| Parallelogram | $A = bh$ |
| Trapezoid | $A = \tfrac{1}{2}h(b_1 + b_2)$ |
| Triangle | $A = \tfrac{1}{2}bh$ |

## VOLUME

| | |
|---|---|
| Prism | $V = bh$ (b = area of base) |
| Pyramid | $V = bh$ (b = area of base) |
| Cube | $V = s^3$ |
| Cylinder | $V = \pi r^2 h$ |
| Cone | $V = 1/3 \, \pi r^2 h$ |
| Sphere | $V = 4/3 \, \pi r^3$ |

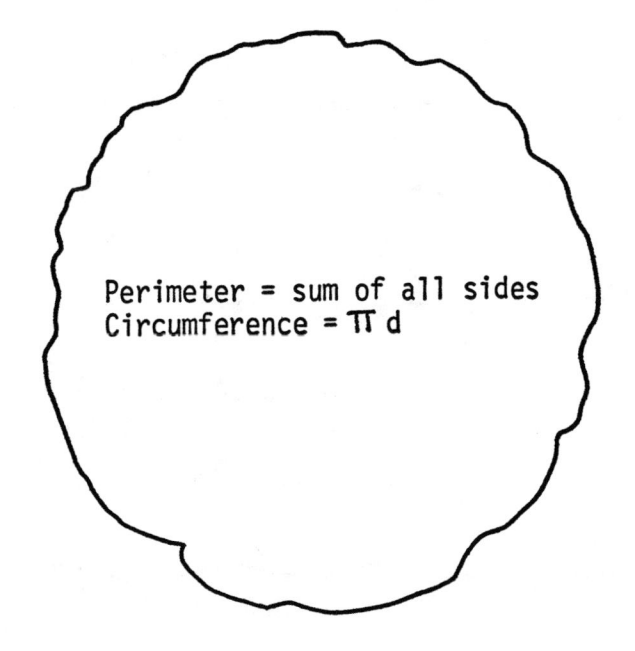

Perimeter = sum of all sides
Circumference = $\pi d$

## SYMBOLS

| | | | |
|---|---|---|---|
| $ | dollar | -3 | negative integer |
| ¢ | cents | $3^4$ | exponential number (3 x 3 x 3 x 3) |
| = | equal | √ | square root |
| ≠ | unequal | + | add |
| < | less than | − | subtract |
| > | greater than | × | multiply |
| { } | empty set | · | multiply |
| ∅ | empty set | ⟌ | divide |
| ∪ | union (of sets) | ÷ | divide |
| ∩ | intersection (of sets) | ⊢⊣ | line segment |
| ≈ | equivalent | → | ray |
| ≉ | unequivalent | ∠ | angle |
| $12_5$ | base 5 numeral | ⌒ | arc |
| 30°F | 30 degrees Farenheit | ≅ | congruent |
| 20°C | 20 degrees Celsius | π | pi (3.14 or 22/7) |
| +9 | positive integer | • | point |

# treasures, tools, and tidbits for math lovers

## TREASURES . . .

*Activities in Mathematics Series.* Glenview, Illinois: Scott Foresman, 1971.

Biggs, Edith, and James MacLean. *Freedom to Learn: An Active Learning Approach to Mathematics.* Reading, Massachusetts: Addison-Wesley, 1972.

Bendick, Jeanne, and Marcia Levin. *Mathematics Illustrated Dictionary.* New York: McGraw, 1965.

Dudney, Henry. *536 Puzzles and Curious Problems.* New York: Scribner's, 1967.

Dumas, Enoch. *Math Activities for Child-Involvement.* Rockleigh, New Jersey: Allyn-Bacon, 1971.

Forte, Imogene, Mary Ann Pangle and Robbie Tupa. *Center Stuff for Nooks, Crannies and Corners.* Nashville, Tennessee: Incentive Publications, 1973.

Gardner, Martin. *The Unexpected Hanging and Other Mathematical Diversions.* New York: Simon and Schuster, 1969.

May, Lola J. *New Math for Adults Only.* New York: Harcourt, Brace and World, 1966.

_____. *Teaching Mathematics in the Elementary School.* New York: MacMillan, 1970.

McWhirter, Norris and Ross. *Guinness Book of World Records.* New York: Sterling, 1973.

Stonaker, Frances. *Famous Mathematicians.* Philadelphia: Lippincott, 1966.

Westcott, Alvin M. and James Smith. *Creative Teaching of Mathematics in the Elementary School.* Boston: Allyn and Bacon, 1967.

TRY THESE GAMES AND PUZZLES:

The Winning Touch
Base Check
Smarty
Sum Fun

from

Ideal School Supply Company
11002 South LaVergne Avenue
Oak Lawn, Illinois   60453

AND THESE TOO:

Krypto
Stamina
Heads Up
Prime Drag
Numble
3-D Tic Tac Toe
Tower Puzzle
Prime Factor

Tuf
Math Match
Moncala
Shotzee
Psych-Out
Soma Puzzle
Fraction Dominoes
Maze Craze

from

Creative Publications
P.O. Box 10328
Palo Alto, California   94303

## TOOLS . . .

multibase arithmetic blocks
Unifix cubes
Chip Trading Program
attribute blocks and games
Invicta math balance
geoboards
tangrams
Geo-Ring Polyhedra Kit
geometric figures and solids
Invicta binary counter
blank playing cards
polyhedra dice
chalkboard gridmaker
chalkboard compass
metric scales
metric rulers and calipers
Invicta simple scales
Ohaus spring scales
Farenheit and Celsius thermometer
Circlemaster compass

These highly recommended tools are all available through:
Creative Publications
P.O. Box 10328
Palo Alto, Calif. 94303

Cuisenaire rods are a _must_ for every classroom.
Write for a catalog:
Cuisenaire Co. of America
12 Church Street
New Rochelle, N.Y. 10805

# TIDBITS . . .

abacus
adding machine
adding machine tape
advertisements
alarm clock
attribute blocks and cards

bags
balance
bank books
beads
beans
blocks
bottle caps
bottles
bowls
boxes
buttons

calculator
calipers
cans
cards
cash register
catalogs
checkers
checks
chips (plastic)
clocks
clothespins
coins
compasses
containers (assorted)
cookbooks
coordinate grids
counting rods
counters
cubes

deposit slips
dice
dot paper

egg cartons
egg timers

flash cards
fraction chart
funnel

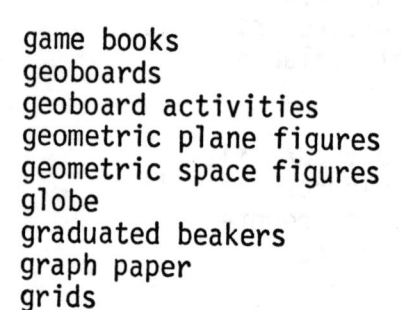

game books
geoboards
geoboard activities
geometric plane figures
geometric space figures
globe
graduated beakers
graph paper
grids

jars
junk box of things to measure and
    compare, i.e., belts
              shoes
              baskets
              nails

liquid measuring devices
   (containers of:  cup
                            pint
                            quart
                            gallon
                            liter sizes)

magazines
maps
marbles
mathematics dictionary
matrix charts
measuring spoons
measuring sticks and tapes
mirrors
multibase blocks

needles
newspapers
number lines

odometers

origami paper

paper money
paper plates
patterns for polyhedra
pegboards
pegs
pictures
pipe cleaners
place value charts
popsicle sticks
posters
protractors
puzzle books
puzzles with:   numbers
                coins
                shapes
                matchsticks
                cards
                words
                designs

recipes
rope
rubber bands
rulers

scales
scissors
scraps of:   flannel
            wallpaper
            fabric
            carpet
            cardboard
            wood

seeds
slates
slide rules
spinners
spoons
stamps
stopwatch
straws
string

tangrams
thermometers
timers
tongue depressors
toothpicks

transparent geoboard
transparent geometric figures
transparent grids

window shades
world atlas

yarn

**bright idea!**

Send everybody on a treasure hunt to gather "math stuff!"

These are some fine spots for searching:

mother's drawers and cupboards
grocery store "back rooms"
laundry rooms
school kitchen
old toy chests and game boxes
basements
soda pop machines
wastebaskets
garage sales
city dumps
principal's office
art room
wallpaper stores

# hung up on a math word?

## (Use this great glossary!)

ADDEND--A number being added in an addition problem.
*In the equation 4 + 7 = 11, 4 and 7 are addends.*

ADDITION--An operation combining two or more numbers.

ADDITIVE INVERSE--For a given number, the number that can be added to give a sum of 0.
*-4 is the additive inverse of + 4 because - 4 + (+4) = 0.*

ALGORITHM--A form commonly used for performing computations involving mathematical operations.

ALTITUDE OF A TRIANGLE--The distance between a point on the base and the vertex of the opposite angle, measured along a line which is perpendicular to the base. (The altitude is also referred to as the height of the triangle.)

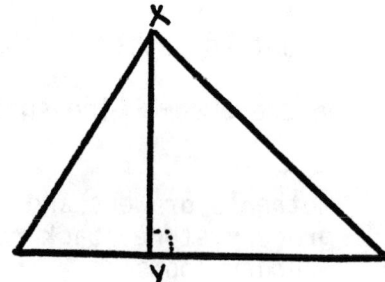

*Segment XY is the altitude in the triangle pictured.*

ANGLE--A figure formed by two rays having a common endpoint (vertex).

An *ACUTE ANGLE* measures less than 90 . (See ∠1 below.)
A *RIGHT ANGLE* measures 90 . (See ∠2 below.)
An *OBTUSE ANGLE* measures more than 90 and less than 180 . (∠3 below.)
A *STRAIGHT ANGLE* measures 180 . (∠4 below.)

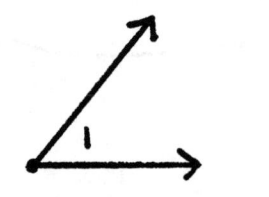

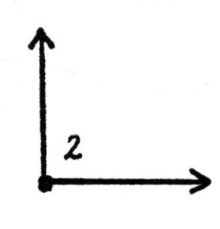

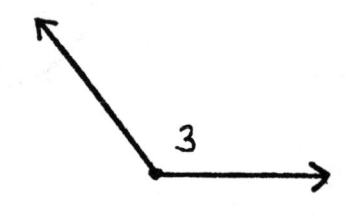

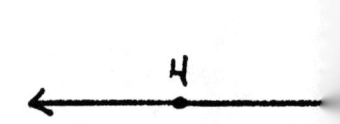

CONGRUENT ANGLES are angles measuring the same.

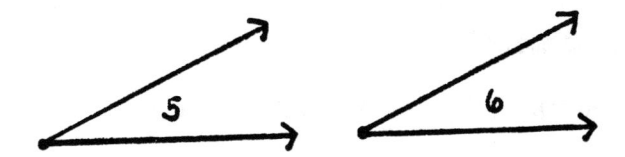

*∠5 is congruent to ∠6 in the illustration above.*

CORRESPONDING ANGLES are formed when a line intersects two parallel lines.
*Corresponding angles are congruent. ∠B and ∠F are corresponding angles in the illustration below.*

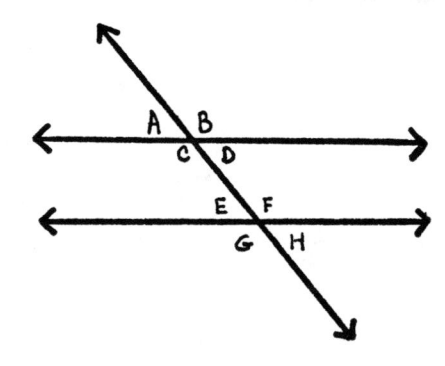

VERTICAL ANGLES are formed opposite one another when two lines intersect.
*Vertical angles are congruent. ∠E and ∠H are vertical angles.*

ARC--A portion of the edge of a circle between any two points on the circle.

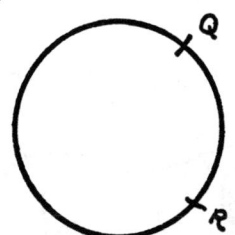

*Segment QR is an arc.*

AREA--The measure of the region inside a closed plane figure. Area is measured in square units.

ASSOCIATIVE PROPERTY FOR ADDITION AND MULTIPLICATION--Rule stating that the grouping of addends or factors does not affect the sum or product.
(3 + 6) + 9 = 3 + (6 + 9)           (2 x 4) x 7 = 2 x (4 x 7)

AVERAGE--The sum of a set of numbers divided by the number of addends.
*The average of 1, 2, 7, 3, 8, and 9 = $\frac{1 + 2 + 7 + 3 + 8 + 9}{6}$ = 5*

AXIS--A number line which may be vertical or horizontal.

BASE-- 1. A side of a geometric figure.

2. Standard grouping of a numeration system

*If a numeration system groups objects by fives, it is called a base 5 system. $23_5$ is a base 5 numeral meaning 2 fives and 3 ones.*

BINARY OPERATION--Any operation involving two numbers.

BISECT--To divide into two congruent parts.

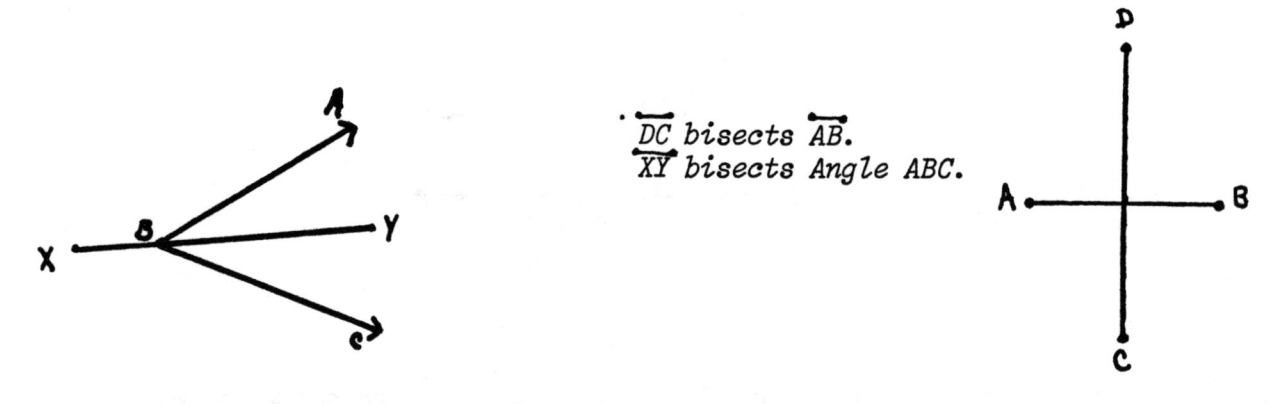

*$\overline{DC}$ bisects $\overline{AB}$.*
*$\overrightarrow{XY}$ bisects Angle ABC.*

CAPACITY--The measure of the amount that a container will hold.

CARDINAL NUMBER--The number of elements in a set.

CARTESIAN SET--A set resulting from the pairing of members of two other sets.

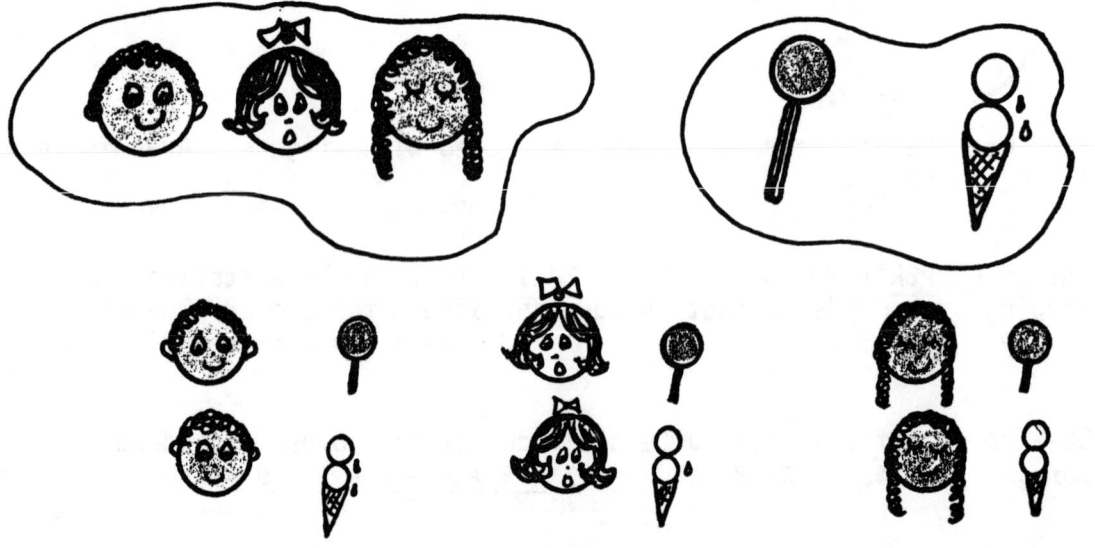

CHANCE--The probability or likelihood of an occurrence.

CHORD--A line segment having endpoints on a circle.

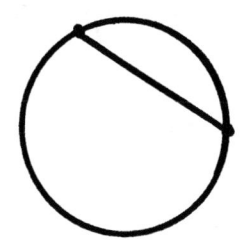

CIRCLE--A closed curve in which all points on the edge are equidistant from a given point in the same plane.

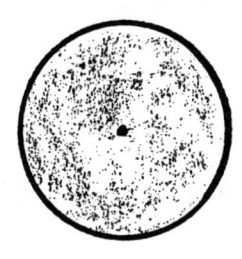

CIRCUMFERENCE--The distance around a circle.
*Circumference = π x diameter.*

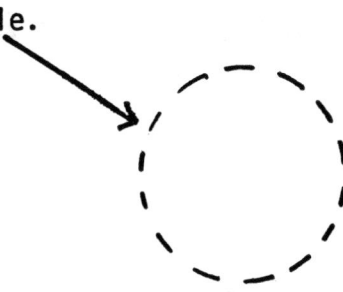

CLOSED FIGURE--A set of points that encloses a region in the same plane; a curve that begins and ends at the same point.

COMMON DENOMINATOR--A whole number that is the denominator for both members of a pair of fractions.
   *3/7 and 5/7, 7 is a common denominator.*

COMMON FACTOR--A whole number which is a factor of two or more numbers.
*3 is a factor common to 6, 9, and 12.*

COMMON MULTIPLE--A whole number that is a multiple of two or more numbers.
*12 is a multiple common to 2, 3, 4, and 6.*

COMMUTATIVE PROPERTY FOR ADDITION AND MULTIPLICATION--Rule stating that the order of addends or factors has no effect on the sum or product.
3 + 9 = 9 + 3   and   4 x 7 = 7 x 4

COMPASS--A tool for drawing circles.

COMPLEX FRACTION--A fraction having a fraction or a mixed numeral as its numerator and/or denominator.
$\frac{3/5}{1/2}$ *is a complex fraction.*

COMPOSITE NUMBER--A number having at least one whole number factor other than 1 and itself.

CONE--A space figure with a circular base.

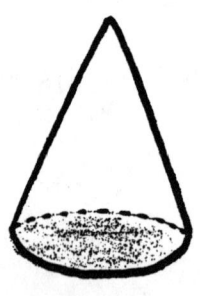

CONGRUENT--Of equal size. The symbol $\cong$ means congruent.

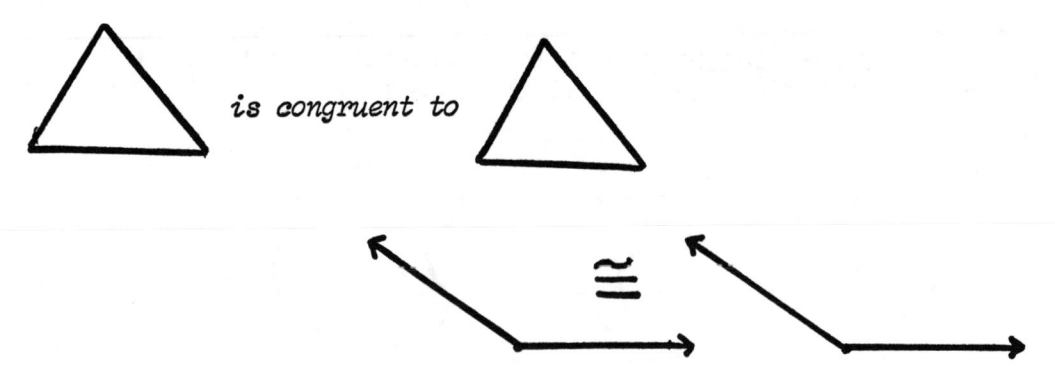

△ *is congruent to* △

⟶ $\cong$ ⟶

COORDINATES--A pair of numbers which gives the location of a point on a plane.

289

CROSS PRODUCT METHOD--Means of testing for equivalent fractions.
If 3/5 ≈ 6/10, then 3 x 10 will equal 5 x 6.

CUBE--A space figure having six congruent square faces.

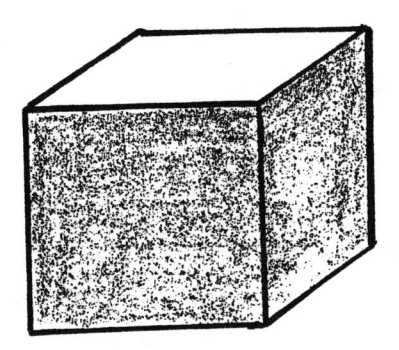

CURVE--A set of points connected by a line segment.

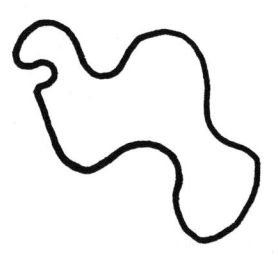

CYLINDER--A space figure having two congruent circular bases.

DATA--A set of scores or information.

DECIMAL NUMERAL--A name for a fractional number expressed with a decimal point, such as .27. 4.03 is a mixed decimal.

DECIMAL SYSTEM--A numeration system based on grouping by tens.

DEGREE--1. A unit of measure used in measuring angles. A circle contains 360 degrees.

2. A unit for measuring temperature.

DENOMINATOR--The bottom number in a fraction. The denominator tells how many parts in all.

**DIAGONAL**--A line segment joining two nonadjacent vertices in a polygon.

*$\overline{AC}$ is a diagonal in the figure below.*

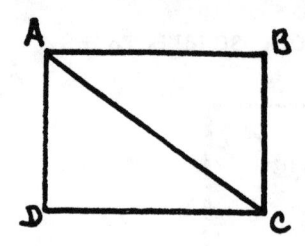

**DIAMETER**--A line segment which has its endpoints on the circle and which passes through the center of the circle.

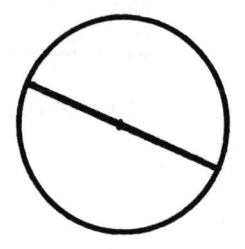

**DIFFERENCE**--The answer in a subtraction problem.

**DIGIT**--A symbol used to write numerals. In the decimal system, there are ten digits: 0-9.

**DISJOINT SETS**--Sets having no members in common.

**DISTRIBUTIVE PROPERTY FOR MULTIPLICATION OVER ADDITION**--Rule stating that when the sum of two or more addends is multiplied by another number, each addend must be multiplied separately, and the products added together.
$$3 \times (4 + 6 + 9) = (3 \times 4) + (3 \times 6) + (3 \times 9)$$

**DIVIDEND**--A number which is to be divided in a division problem.

**DIVISIBILITY**--A number is divisible by a given number if the quotient is a whole number.

*189 is divisible by 9 because 189 ÷ 9 is a whole number.*

**DIVISION**--The operation of finding a missing factor when the product and one factor are known.

**DIVISOR**--The factor used in a division problem for the purpose of finding the missing factor.
*In the following example, 23 is the divisor:* 23 ⟌ 1980

291

EDGE--A line segment formed by the intersection of two faces of a geometric space figure.

ELEMENTS--The members of a set.

EMPTY SET--A set having no elements; also called a null set.
*An empty set is represented like this:* { }   *or*   ∅

ENDPOINT--A point at the end of a line segment or ray.

EQUATION--A mathematical sentence which states that two expressions are equal.
*7 x 9 = 3 + (4 x 15) is an equation.*

EQUATOR--An imaginary line at 0 degrees latitude on the earth's grid.

EQUILATERAL--Having sides of the same length

*Figure ABC is an equilateral triangle. All of its sides are the same length.*

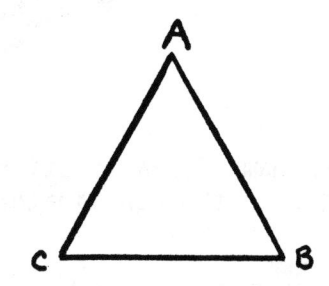

EQUIVALENT FRACTIONS--Fractions that name the same fractional number.

EQUIVALENT SETS--Sets having the same number of members.

EVEN NUMBER--One of the set of whole numbers having 2 as a factor.

EXPANDED NOTATION--Method of writing a numeral to show the value of each digit.
*5327 = 5000 + 300 + 20 + 7*

EXPONENT--A numeral telling how many times a number is to be used as a factor.
*In $6^3$, the exponent is 3. $6^3$ means 6 x 6 x 6, or 216.*

FACE--A plane region serving as a side of a space figure.
*This cube has six faces.*

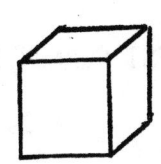

FACTOR--One of two or more numbers that can be multiplied to find a product.
*In the equation 6 x 9 = 54, 6 and 9 are factors.*

FINITE SET--A set having a specific number of elements.
*2, 5, 9, 15 is a finite set.*

FLIP--To turn over a geometric figure. The size or shape of the figure does not change.

FRACTION--A name for a fractional number written in the form a/b. A is the numerator; b is the denominator.

FRACTIONAL NUMBER--A number that can be named as a fraction a/b with a and b being any numbers, except b cannot be zero.

FREQUENCY GRAPH--A way to organize and picture data using a grid.

FUNCTION: A set of ordered pairs of numbers which follow a function rule, and in which no two first numbers are the same.
*{(2,5) (3,6) (4,7) (5,8) (6,9)} The rule for this set: add 3.*

GEOMETRY--The study of space and figures in space.

GRAM--A standard unit for measuring weight in the metric system.

GRAPH--A drawing showing relationships between sets of numbers.

GREATEST COMMON FACTOR--The largest number that is a factor of two other numbers.
*6 is the greatest common factor of 18 and 24.*

GRID--A set of horizontal and vertical lines spaced uniformly.

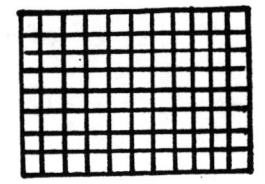

HEMISPHERE--Half a sphere.

HORIZONTAL--A line that runs parallel to a base line.

HYPOTENUSE--The longest side of a right triangle located opposite the right angle.

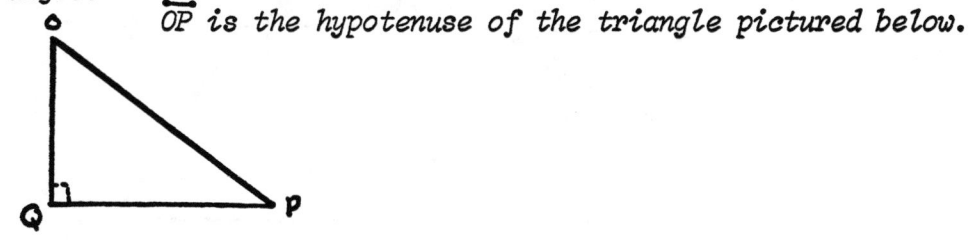

$\overline{OP}$ *is the hypotenuse of the triangle pictured below.*

IDENTITY ELEMENT FOR ADDITION--Zero is the identity element for addition, because any number plus 0 = that number. (3 + 0 = 3)

IDENTITY ELEMENT FOR MULTIPLICATION--One is the identity element for multiplication, because any number x 1 = that number. (17 x 1 = 17)

IMPROPER FRACTION--A fraction having a numerator equal to or greater than the denominator, therefore naming a number of one or more.
*7/4 is an improper fraction.*

INEQUALITY--A number sentence showing that two groups of numerals stand for different numbers.
*The signs* <, >, ≠ *show inequality.  7 + 5 > 12 - 9*

INFINITE SET--A set having an unlimited number of members.

INTEGER--Any member of the set of positive or negative counting numbers and zero.
$$\{\ldots -4, -3, -2, -1, 0, 1, 2, 3, 4, \ldots\}$$

INTERSECTION OF LINES--The point at which two lines meet.

INTERSECTION OF PLANES--A line formed by the set of points at which two planes meet.

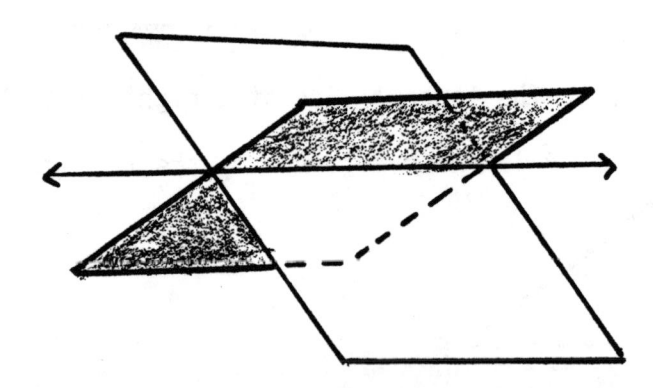

INTERSECTION OF SETS--The set of members common to each of two or more sets.
*The intersection of A and B below is $\{3, 7, 8\}$.*
*The intersection of B and C below is $\{\ \}$*
*The symbol $\cap$ stands for intersection.*

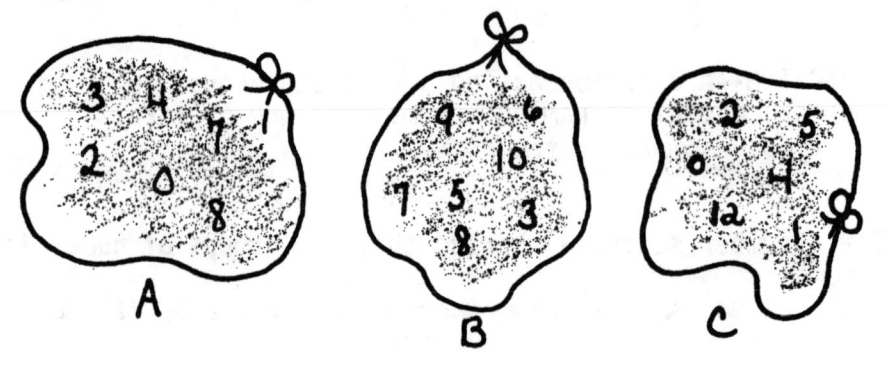

INVERSE--Opposite. Addition and subtraction are inverse operations. Multiplication is the inverse of division.

295

LATERAL FACES--The plane surfaces of a space figure which are not bases. The lateral faces of the triangular prism below are shaded.

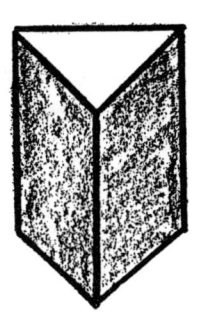

LATITUDE--The distance north or south of the equator, measured in degrees. Lines of latitude run parallel to the equator.

LEAST COMMON DENOMINATOR--The smallest whole number which is a multiple of the denominators of two or more fractions.
*12 is the least common denominator for 1/3 and 3/4.*

LEAST COMMON MULTIPLE--The smallest whole number which is divisible by each of two or more given numbers.
*18 is the least common multiple of 2, 6, 9, and 18.*

LIKE FRACTIONS--Fractions having the same denominator.
*2/9 and 12/9 are like fractions.*

LINE--A set of points along a path in a plane.

LINE SEGMENT--Part of a line consisting of a path between two endpoints.
*$\overline{AB}$ and $\overline{CD}$ below are line segments.*

LINEAR MEASURE (or length)--The measure of distance between two points along a line.

LITER--A metric system unit of measurement for liquid capacity.

LOGIC--Principles of reasoning.

LONGITUDE--The distance east or west of the prime meridian, measured in degrees.  Lines of longitude run north and south on the earth's grid, meeting at the poles.


LOWEST TERMS--For a fraction, having a numerator and denominator with no common factor greater than 1.
    *7/9 is a fraction in lowest terms.*


MEAN--Average:  the sum of numbers in a set divided by the number of addends.
    *The mean of {6, 8, 9, 19, 38} is 80/5 or 16.*


MEASUREMENT--The process of finding the length, area, capacity, or amount of something.


MEDIAN--The middle number in a set of numbers.  The median is determined by arranging numbers in order from lowest to highest, then counting to the middle.
    *The median of {3, 8, 12, 17, 20, 23, 27} is 17.*


METER--A metric system unit of linear measurement.


METRIC SYSTEM--A system of measurement based on the decimal system.


MIXED NUMERAL--A numeral that includes a whole number and a fractional number or a whole number and a decimal.
    *7½ and 37.016 are mixed numerals.*


MODE--The score or number found most frequently in a set of numbers.


MODULAR NUMBER SYSTEM--A number system that uses a limited number of units for counting.

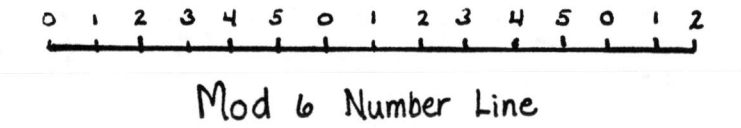

$$\begin{array}{c|c|c|c|c|c|c|c|c|c|c|c|c|c|c|c|c|c|c}0&1&2&3&4&5&0&1&2&3&4&5&0&1&2\end{array}$$

    Mod 6 Number Line


MULTIPLE--The product of a whole number and any other whole number.
    *12 is a multiple of 3 because 3 x 4 = 12.*


MULTIPLICATION--An operation involving repeated addition.
    *4 x 5 means 4 + 4 + 4 + 4 + 4.*

MULTIPLICATIVE INVERSE--For any given number, the number that will yield a product of 1.
> *4/3 is the multiplicative inverse of 3/4 because 4/3 x 3/4 = 1.*

NAPIER'S BONES--A set of sticks or rods bearing multiplication facts.

NEGATIVE INTEGER--One of a set of counting numbers that is less than 0.

NON-DECIMAL NUMERALS--Numerals representing numbers in a system having a base other than 10.
> *$23_4$ is a base 4 numeral.*

NUMBER--A mathematical idea concerning the amount contained in a set.

NUMBER LINE--A line which has numbers corresponding to points along it.

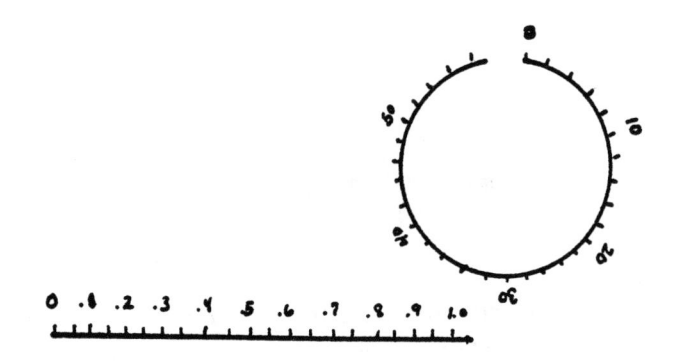

NUMERAL--A symbol used to represent or name a number.

NUMERATION SYSTEM--A system of symbols used to express numbers.

NUMERATOR--The number above the line in a fraction.

ODD NUMBER--A whole number belonging to the set of numbers equal to (nx2) + 1.
> *{1, 3, 5, 7, 9, . . .} are odd numbers.*

ORDERED PAIR--A pair of numbers in a certain order, the <u>order</u> being of significance.

ORDINAL NUMBER--A number telling the place of an item in an ordered set. Sixth describes an ordinal number.

ORIGIN--The beginning point on a number line. The origin is often zero.

298

PALINDROME--A number which reads the same forward and backward.
*343 and 87678 are palindromic numbers.*

PARALLEL LINES--Lines in the same plane which do not intersect.

PARALLELOGRAM--A quadrilateral whose opposite sides are parallel.

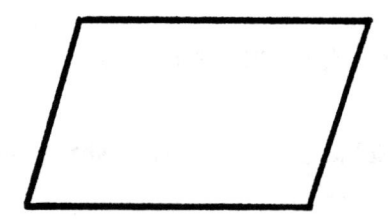

PERCENT--A comparison of a number with 100.
*43 compared to 100 is 43%.*

PERIMETER--The distance around the outside of a closed figure.
*The perimeter of Figure A is 6 cm.*

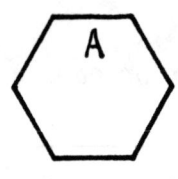

PERIODS--Groups of three digits in numbers.

*723, 301, 611*
*millions period   thousands period   units period*

PERPENDICULAR LINES--Two lines in the same plane that intersect at right angles.

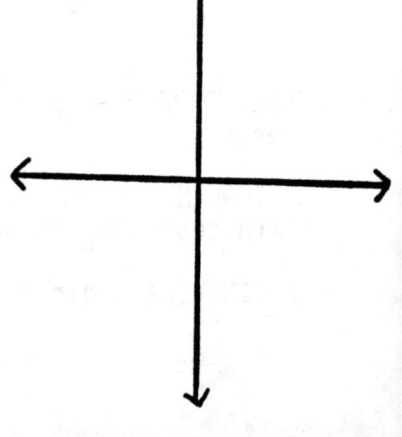

PLACE VALUE--The value assigned to a digit due to its position in a numeral.

PLANE--The set of all points on a flat surface which extends indefinitely in all directions.

PLANE FIGURE--A set of points in the same plane enclosing a region.
*Figures A and B are plane figures.*

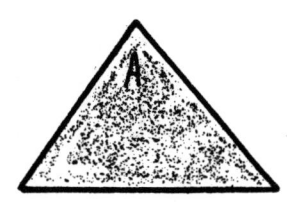

POINT--An exact location in space.

POLYGON--A simple closed plane figure having line segments as sides.

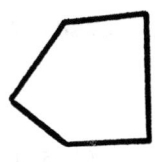

POLYHEDRON--A space figure formed by intersecting plane surfaced called faces.

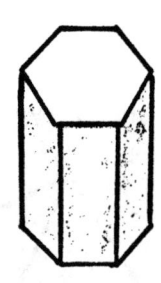

POSITIVE INTEGER--One of a set of counting numbers that is greater than zero.

PRIME FACTOR--A factor that is a prime number.
*1, 2, and 5 are prime factors of 20.*

PRIME MERIDIAN--An imaginary line on the earth's grid located at 0° longitude, running north and south through Greenwich, England.

PRIME NUMBER--A number that has as its only whole number factors one and itself.
*2, 3, 7, 11 . . . are prime numbers.*

PRISM--A space figure with two parallel congruent polygonal faces (called bases). The prism is named by the shape of its bases.
*Figure A is a triangular prism. Figure B is a rectangular prism.*

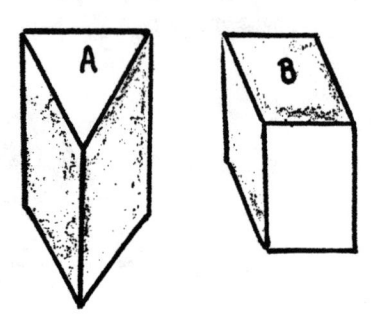

PROBABILITY--A study of the likelihood that an event will occur.

PRODUCT--The answer in a multiplication problem.

PROPORTION--A number statement of equality between two ratios.
*Example: 3/7 = 9/27*

PROTRACTOR--An instrument used for measuring angles.

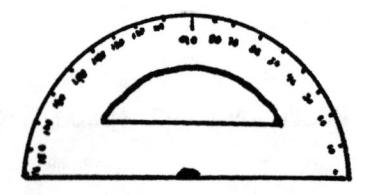

PYRAMID--A space figure having one polygonal base and triangular faces which have a common vertex.

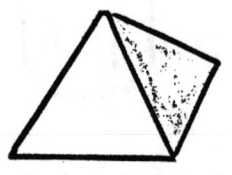

PYTHAGOREAN THEOREM--A proposition stating that, in a right triangle, the sum of the squares of the two shorter sides is equal to the square of the third side.

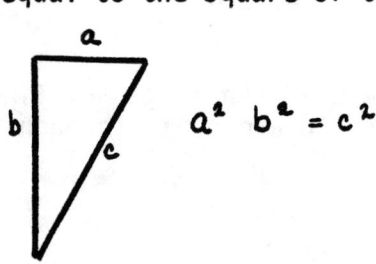

$a^2 \ b^2 = c^2$

QUADRILATERAL--A four-sided polygon.

QUOTIENT--The answer in a division problem.

RADIUS--A line segment having one endpoint in the center of the circle and the other on the circle.

RATE--A comparison of two quantities.

RATIO--A comparison of two numbers expressed as a/b, meaning a ÷ b.

RAY--A portion of a line extending from one endpoint indefinitely in one direction.

RECIPROCALS--A pair of numbers whose product is one.
*1/2 and 2/1 are reciprocals.*

RECIPROCAL METHOD FOR DIVIDING FRACTIONS--A means of dividing fractions that involves replacing the divisor with its reciprocal. and multiplying.

*2/3   4/7 = 2/3 x 7/4 = 14/12 = 1 1/6*

RECTANGLE--A parallelogram having four right angles.

302

REGION--The set of all points on a closed curve and its interior.

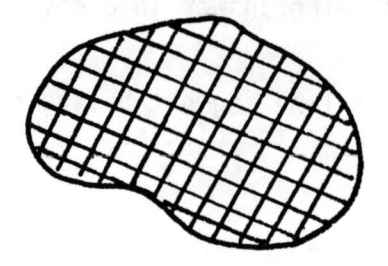

RELATION--A set of ordered pairs.

RENAME (or regroup)--To name numbers with a different set of numerals.

REPLACEMENT SET--A set of numbers which could replace a variable in a number sentence.

RHOMBUS--A parallelogram having congruent sides.
*Figures A and B below are rhombuses.*

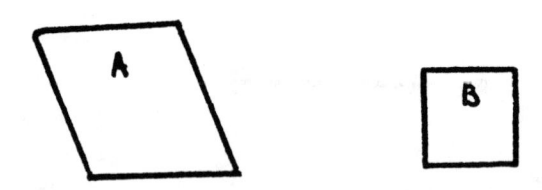

SEQUENCE--A continuous series of numbers ordered according to a pattern.

SET--A collection of items (called members or elements).

SIMILARITY--A property of geometric figures having angles of the same size.
*Angles X and Y are similar in the illustration below.*
*Figures A and B are similar triangles.*

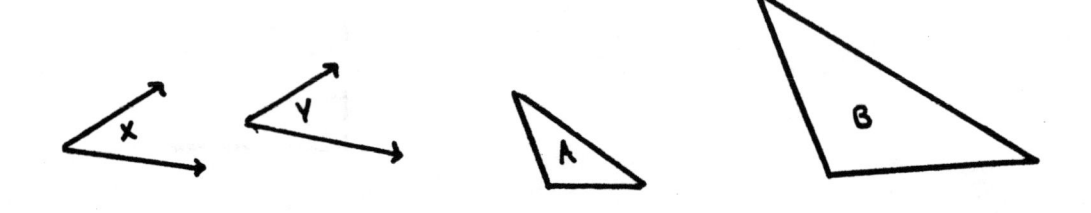

SIMPLE CLOSED CURVE OR FIGURE--A closed curve whose path does not intersect itself.

*The figures below are simple closed curves.*

SKEW LINES--Lines that are not in the same plane and do not intersect.

SLIDE--Moving a figure without turning or flipping it. The shape or size of a figure is not changed by a slide.

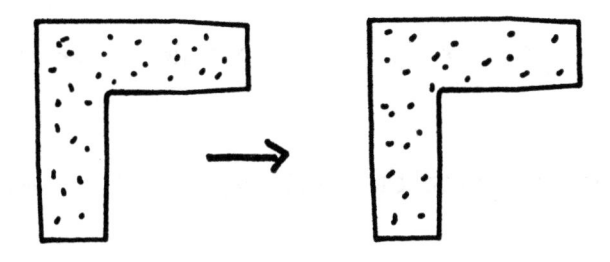

SOLUTION SET--The set of possible solutions for a number sentence.

SPACE FIGURE--A figure which consists of a set of points lying in two or more planes.

SPHERE--A space figure formed by a set of points lying equidistant from a center point.

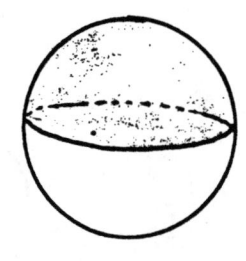

SQUARE--A rectangle with all sides congruent.

SQUARE ROOT--A number, which when multiplied by itself, yields a given product.

$\sqrt{25}$ = 5 because 5 x 5 = 25.

SUBSET--Every member of a set, or any combination of the members of a set.

SUBTRACTION--The operation of finding a missing addend when one addend and the sum are known.

SUM--The answer in an addition problem resulting from the combination of two addends.

SURFACE--A region lying on one plane.

SURFACE AREA--The space covered by a plane region or by the faces of a space figure.

SYMMETRIC FIGURE--A figure having two halves that are reflections of one another. A line of symmetry divides the figure into two congruent parts.
*The figures illustrated below are symmetric.*

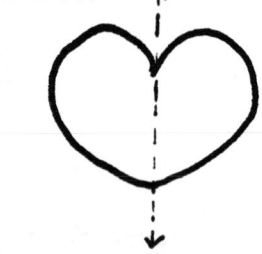

TANGENT--A line which touches a figure at one point and has all its other points outside the figure.

*Line GH is tangent to the circle at Point X.*

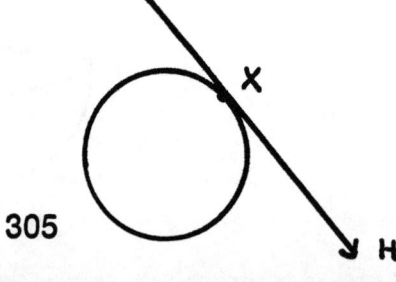

305

TERMS OF A FRACTION--The numerator and the denominator of a fraction.

TRAPEZOID--A quadrilateral having only two parallel sides.

TRIANGLE--A three-sided polygon.

  *ACUTE TRIANGLE*--A triangle in which all three angles are less than 90°.

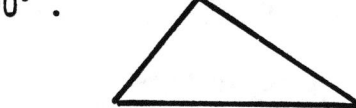

  *ISOSCELES TRIANGLE*--A triangle with at least two congruent sides.

  *OBTUSE TRIANGLE*--A triangle having one angle greater than 90°.

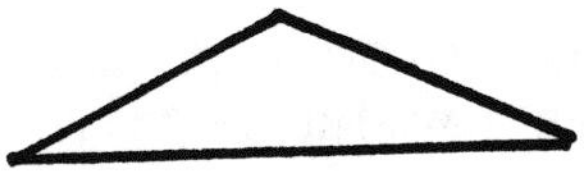

  *RIGHT TRIANGLE*--A triangle having one 90° angle.

  *SCALENE TRIANGLE*--A triangle in which no two sides are congruent.

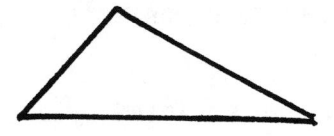

TURN--A move in geometry which involves turning but not flipping a figure. The size or shape of the figure is not changed by a turn.

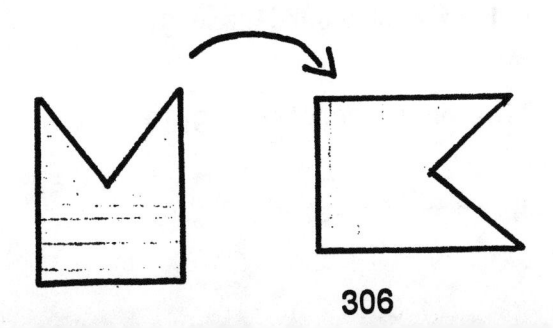

UNION OF SETS--A set containing the combined members of two or more sets.
　　　*The union of Sets A and B below is {7, 12, 14, 20, 26, 20}.*
　　　*The symbol U represents union.*

UNIT--1.　The first whole number.

　　　2.　A determined quantity used as a standard for measurement.

VARIABLE--A symbol in a number sentence which could be replaced by a number.
　　　*In 3 + 9x = 903, x is the variable.*

VENN DIAGRAM--A pictorial means of representing sets and the union or intersection of sets.

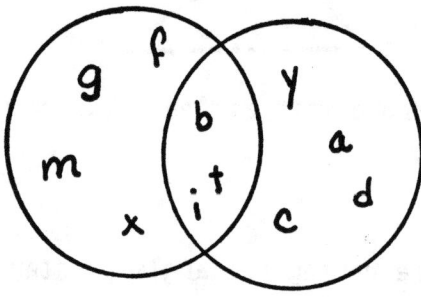

VERTEX--A common endpoint of two rays forming an angle, two line segments forming sides of a polygon or two planes forming a polyhedron

VERTICAL--A line that is perpendicular to a horizontal base line.

VOLUME--The measure of capacity or space enclosed by a space figure.

WHOLE NUMBER--A member of the set of numbers 0, 1, 2, 3, 4, . . . .

X-AXIS--The horizontal number line on a coordinate grid.

Y-AXIS--The vertical number line on a coordinate grid.

ZERO--The number of members in an empty set.

372.86
F851k

88994